# Math Mammoth Grade 5
# Skills Review Workbook
# Answer Key

*By Maria Miller*

# Contents

# Chapter 1: The Four Operations

## Skills Review 1, p. 7

1. a. 920, 850, 780, 710, 640, 570, 500, 430, 360
   b. 540, 750, 960, 1170, 1380, 1590, 1800, 2010, 2220

2. a. 8   b. 6   c. 2

3.
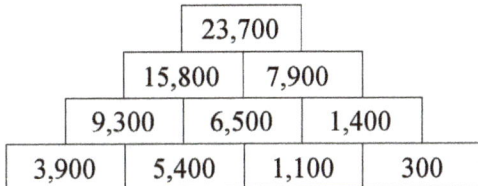

4. a. 960   b. 270   c. 420   d. 2,600

5. She originally had $18 + $26 + 49 = $93.

6. They earned $28 + (2 × $28) = $84 in total.

7. $9,348 - 3,762 = x$  OR  $9,348 - x = 3,762$
   $x + 3,762 = 9,348$
   Solution: $x = 5,586$

## Skills Review 2, p. 8

1. a. $75 \times 9 = 675$
   b. The expression is 18,000 ÷ 600. If we want to find its value (which was not asked but many students may do), we would get 30.

2. $137 \div 6 = 22$ R5  He will need 23 boxes.

3. a. 12   b. 44   c. 76   d. 53

4. a. $(23 - 9) + 37$ OR $37 + (23 - 9)$
   b. $200 - (6 + 70)$

5. a. Bruce drives 26 more miles round-trip than Forrest.
   b. Aiden drives 28 fewer miles round-trip than Mia.
   c. Claire drives 420 miles round-trip during a five-day workweek.

6. a. 68   b. 114   c. 84

## Skills Review 3, p. 9

1.

a. $9 \times 37 = 9 \times 30 + 9 \times 7$

   $= 270 + 63 = 333$

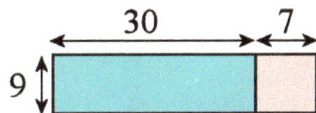

b. $4 \times (6 + 10)$

   $= 4 \times 6 + 4 \times 10$

   $= 24 + 40$

   $= 64$

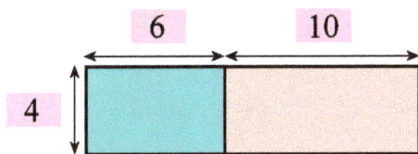

## Skills Review 3, cont.

**2.**

| | | | | | |
|---|---|---|---|---|---|
| a. 2 | | | | | |
| 2 | | b. 6 | | b. 1 | e. 8 |
| a. 5 | 3 | 1 | | | 0 |
| | | 2 | | c. 2 | 0 |
| | | d. 5 | d. 9 | 0 | |
| | | | 3 | | f. 5 |
| | | e. 8 | 4 | 0 | 0 |
| | | | | | 0 |

**3.** a. equation   b. expression   c. equation

**4.**

| + 600 | − 800 | + 600 | ÷ 40 | × 20 | − 700 |
|---|---|---|---|---|---|
| 2800 → 3400 | → 2600 | → 3200 | → 80 | → 1600 | → 900 |

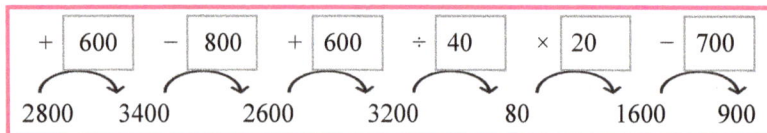

## Skills Review 4, p. 10

**1.** Estimates will vary.
  a. Estimate: $7 \times 5{,}000 = 35{,}000$
    Exact: $7 \times 4{,}652 = 32{,}564$
  b. Estimate: $9 \times 80{,}000 = 720{,}000$
    Exact: $9 \times 81{,}739 = 735{,}651$

**2.**

| | | | |
|---|---|---|---|
| 248,000 | 310,000 | 228,000 | 284,000 |
| 123,000 | 114,000 | 150,000 | 141,000 |
| 57,000 | 60,500 | 74,000 | 119,500 |
| 30,200 | 28,500 | 36,900 | 58,600 |
| 14,175 | 15,125 | 14,250 | 18,400 |
| 7,060 | 8,010 | 7,120 | 7,125 |

**3.** a. 13,500   b. 531   c. 11,300   d. 264

**4.** He paid $2,940 in total.

**5.** Michael paid $5 \times \$60 \div 3 = \$100$.

**6.** (3) $(2 \times \$5 + \$26) \div 3 = \$12$. Joy's share was $12.

## Skills Review 5, p. 11

1. a. 717,990     b. 382,872     c. 948,898

2. a. M = 60     b. M = 54     c. M = 12

3. a. $34 + s = 150$     b. $y - 57$

4. 2050, 1900, 1750, 1600, 1450, 1300, 1150, 1000, 850

5.

| 250 | 250 | 250 | 250 | 250 | 250 |
|-----|-----|-----|-----|-----|-----|

$\longleftarrow$ ———— R ———— $\longrightarrow$

R = 1,500

## Skills Review 6, p. 12

1. a. 609     b. 804     c. 910

2. Estimates may vary. Please check the student's answer.
   Estimate: $7,000 - (40 \times \$30 + 20 \times \$60) = \$4,600$
   Exact: $7,000 - (35 \times \$29 + 18 \times \$56) = \$4,977$

3. a. $228.15     b. $289.74     c. $1,116.64     d. $3,626.92

4. Each one received $469.

5. Eric collected $3 \times 48 = 144$ cans, and Layla collected $144 \div 2 = 72$ cans.
   In total, they collected $48 + 144 + 72 = 264$

6. a. 150,000     b. 8,400     c. 33,000     d. 160,000

## Skills Review 7, p. 13

1.

| Table of 16: | | Check: |
|---|---|---|
| $2 \times 16 = 32$ | | |
| $3 \times 16 = 48$ | | 4 1 5 |
| $4 \times 16 = 64$ | | × 1 6 |
| $5 \times 16 = 80$ | | |
| $6 \times 16 = 96$ | | 2 4 9 0 |
| $7 \times 16 = 112$ | | + 4 1 5 0 |
| $8 \times 16 = 128$ | | |
| $9 \times 16 = 144$ | | 6 6 4 0 |

```
        4 1 5
16)6 6 4 0
   6 4
     2 4
     1 6
         8 0
         8 0
           0
```

2. $540 \div N = 6$; N = 90

3. a. Kevin received $6.75 in change. $20 - (\$3.50 + 4 \times \$1.25 + \$4.75) = \$6.75$.
   b. Marlene bought 6 towels. $(\$15 - 2) \times 6 = \$78$ or $78 \div (\$15 - \$2) = 6$

## Skills Review 7, cont.

**4.**

$14 \times (5 + 7 + 9)$

$= 14 \times 5 + 14 \times 7 + 14 \times 9$

$= 70 + 98 + 126$

$= 294$

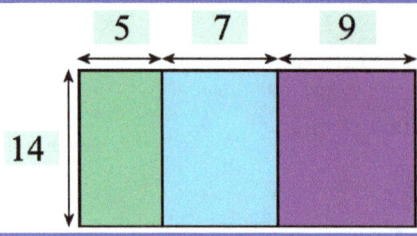

## Skills Review 8, p. 14

1. a. $s = 80$   b. $x = 42$   c. $y = 410$

2.

| | | |
|---|---|---|
| $2 \times 52 = 104$ | $\phantom{00}1\,6\,4$ | $\phantom{000}1\,6\,4$ |
| $3 \times 52 = 156$ | $52\,\overline{)8\,5\,7\,3}$ | $\times\phantom{00}5\,2$ |
| $4 \times 52 = 208$ | $\underline{-5\,2}$ | |
| $5 \times 52 = 260$ | $\phantom{0}3\,3\,7$ | $\phantom{00}3\,2\,8$ |
| $6 \times 52 = 312$ | $\underline{-3\,1\,2}$ | $\underline{+8\,2\,0\,0}$ |
| $7 \times 52 = 364$ | $\phantom{00}2\,5\,3$ | $\phantom{0}8\,5\,2\,8$ |
| $8 \times 52 = 416$ | $\underline{-\phantom{0}2\,0\,8}$ | $+\phantom{00}4\,5$ |
| $9 \times 52 = 468$ | $\phantom{000}4\,5$ | $\phantom{0}\overline{8\,5\,7\,3}$ |

3. Mom had **8** apples left.

4. Estimates will vary. Please check the student's estimates.
   a. Estimate: $4 \times \$30 = \$120$       Exact: $4 \times \$27.45 = \$109.80$
   b. Estimate: $\$60 \times 12 = \$720$       Exact: $\$58.60 \times 12 = \$703.20$

## Skills Review 9, p. 15

1. $940 - 510 = 2x$   OR   $940 - 2x = 510$
   $2x + 510 = 940$
   Solution: $x = 215$

2.

Bag 858 plums, 3 plums in each bag.

| Plums | Bags | | |
|---|---|---|---|
| $8\,5\,8$ | | $\phantom{0}2\,8\,6$ | |
| $-\phantom{0}6\,0\,0$ | $2\,0\,0$ | $3\,\overline{)8\,5\,8}$ | |
| | | $\underline{-6}$ | |
| $\phantom{0}2\,5\,8$ | | $\phantom{0}2\,5$ | |
| $-\phantom{0}2\,4\,0$ | $8\,0$ | $\underline{-2\,4}$ | |
| | | $\phantom{00}1\,8$ | |
| $\phantom{00}1\,8$ | | $\underline{-\phantom{0}1\,8}$ | |
| $-\phantom{00}1\,8$ | $6$ | $\phantom{00}0$ | |
| $\phantom{000}0$ | $2\,8\,6$ | | |

3. a. 39    b. 72    c. 77    d. 204    e. 60    f. 222

4. She kept 4 gal $-$ 1 1/2 gal = 2 1/2 gal. Each gallon is 4 quarts, so 2 1/2 gal makes $4 + 4 + 2 = 10$ quarts. She filled 10 jars.

5. a. 646,000    b. 920,000    c. 340,000

## Skills Review 10, p. 16

1. a. Yes, because 90 divided by 6 is an even division.
   b. No, because 330 divided by 8 leaves a remainder (is not an even division).

2.

a.
```
          1 3 9      Check:
  4 9)6 8 2 3          1 3 9
      4 9            ×   4 9
      1 9 2          1 2 5 1
      1 4 7        + 5 5 6 0
          4 5 3      6 8 1 1
          4 4 1    +     1 2
            1 2      6 8 2 3
```

b.
```
          7 9 2      Check:
    6)4 7 5 2          7 9 2
      4 2            ×     6
        5 5          4 7 5 2
        5 4
          1 2
          1 2
            0
```

3.

| a. $108 = (8 + 4) \times 9$ | b. $7 \times 6 = 9 \times 4 + 6$ | c. $5 + 9 = (80 - 24) \div 4$ |
|---|---|---|

4. Dad gave Carl $16 + 4 \times \$2.40 + \$14.40 = \$40$.

5.

| 120 | 120 | 120 | 120 | 120 | 120 | 120 | 120 | 120 |
|---|---|---|---|---|---|---|---|---|

← ————————— R ————————— →

R = 1,080

## Skills Review 11, p. 17

1. a. 1, 2, 4, 7, 14, 28
   b. 1, 2, 3, 6, 7, 14, 21, 42

2.

Bag 857 lemons, 6 lemons in each bag.

```
   Lemons        Bags            1 4 2
                            6)8 5 7
     8 5 7                    - 6
   - 6 0 0      1 0 0          2 5
   -------                   - 2 4
     2 5 7                      1 7
   - 2 4 0       4 0          - 1 2
   -------                       5
     1 7
 -   1 2         2
   -------
       5       1 4 2
```

3. a. $x - 14$    b. $40 \div b = 8$    c. $y + 7 + 9 = 37$

4. a. 105    b. 48    c. 22,140

5. a. 941,112    b. 473,770    c. 994,004

9

1.

| Divisible by | 2 | 3 | 4 | 5 | 6 | 9 |
|---|---|---|---|---|---|---|
| 2,932 | X | | X | | | |
| 453 | | X | | | | |
| 7,000 | X | | X | X | | |
| 8,715 | | X | | X | | |

| Divisible by | 2 | 3 | 4 | 5 | 6 | 9 |
|---|---|---|---|---|---|---|
| 4,218 | X | X | | | X | |
| 729 | | X | | | | X |
| 864 | X | X | X | | X | X |
| 5,367 | | X | | | | |

2. They each paid $5,871 ÷ 3 = $1,957.

3.

49 + 7    90 − 18
200 ÷ 8    12 × 200
62 − 5    38 + 18
3 × 6 × 4    1000 ÷ 40
3000 ÷ 60    83 − 26
76 + 9    7 × 40
80 × 30    600 ÷ 12
400 − 120    130 − 45

4. a. false    b. true    c. false    d. false    When changing one number in (a), (c), and (d) answers will vary, as there are several possibilities. For example:

a. $9 + \dfrac{49}{7} = 16$;   OR   $9 + \dfrac{42}{7} = 15$;   OR   $8 + \dfrac{49}{7} = 15$

c. $4 \times 2 = 24 \div 6 + 4$;   OR   $4 \times 2 = 18 \div 6 + 5$;   OR   $4 \times 2 = 24 \div 8 + 5$

d. $15 - 7 = 20 \div 4 + 3$;   OR   $15 - 7 = 20 \div 10 + 6$;   OR   $15 - 4 = 20 \div 4 + 6$

5. a. 525    b. 160    c. 610

6. a. $96 = 2 \times 2 \times 2 \times 2 \times 2 \times 3$    b. $85 = 5 \times 17$    c. $77 = 7 \times 11$

# Chapter 2: Large Numbers and the Calculator

## Skills Review 13, p. 19

1. a. 62 R4    b. 40 R55

2.

| 10 | 7 | 3 | 5 |
|----|----|----|----|
| 15 | 12 | 9 | 17 |
| 8 | 14 | 5 | 13 |
| 7 | 6 | 11 | 9 |
| 11 | 18 | 4 | 19 |
| 22 | 16 | 20 | 10 |

3. a. 400    b. 28,000    c. 1,800,000    d. 160,000

4. Lauren and Emma drove $1,368 - (480 + 395) = 493$ miles on the third day.

5. a. $8 \times y = 56$;  $y = 7$
   b. $y \div 8 = 56$;  $y = 448$

## Skills Review 14, p. 20

1. c. $562 - (3 \times \$28 + \$94)$

2. a. 636,400    b. 315,000    c. 667,200

3. Andrew is 1 ft 7 in taller than Brenda.

4. a. 5 R3;  5 R4;  5 R5
   b. 8 R4;  8 R5;  8 R6
   c. 7 R1;  7 R2;  7 R3

5. a. >    b. <    c. <

6. a. 72    b. 62    c. 3,950

## Skills Review 15, p. 21

1. a. $s = 150$    b. $x = 54$    c. $y = 80$

2. a. 9; 18    b. 15; 30    c. 62; 124

3.

| a. 875,<u>3</u>12,461 | b. <u>9</u>,503,794,825 |
|---|---|
| Place: <u>hundred thousands place</u> | Place: <u>billions place</u> |
| Value: <u>300,000 (three hundred thousand)</u> | Value: <u>9,000,000,000 (nine billion)</u> |

4. Vanessa can sew buttons onto 52 shirts. She will have 5 buttons left over.

5. a. 78    b. 198    c. 680

Mystery Number: 154

**Skills Review 16, p. 22**

1.

$11 \times (5 + 7 + 10)$

$= 11 \times 5 + 11 \times 7 + 11 \times 10$

$= 55 + 77 + 110$

$= 242$

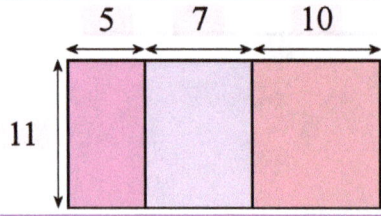

2.

Bag 1,960 apples, 8 apples in each bag.

| Apples | Bags |
|--------|------|
| 1,960 | |
| − 1,600 | 200 |
| 360 | |
| − 320 | 40 |
| 40 | |
| − 40 | 5 |
| 0 | 245 |

```
         2 4 5
    8 )1 9 6 0
      -1 6
        3 6
       -3 2
          4 0
        - 4 0
            0
```

3. a. 23,330   b. 271   c. 21,270   d. 514

4. Her brothers ate $3 \times 3/4$ C = 2 1/4 C. Then Audrey ate $4 - 2 1/4$ C = 1 3/4 cups of ice cream.

5. No, Gregory did not receive the correct change. He should have received $20 − 8 × \$1.48 = \$8.16$ in change.

6.

| | |
|---|---|
| a. $5 \times 5 \times 5 \times 5 \times 5 = 5^5 = 3{,}125$ | d. $2 \times 2 \times 2 \times 2 \times 2 \times 2 = 2^6 = 64$ |
| b. $10 \times 10 \times 10 \times 10 = 10^4 = 10{,}000$ | e. $9 \times 9 \times 9 = 9^3 = 729$ |
| c. seven squared $= 7^2 = 49$ | f. four cubed $= 4^3 = 64$ |

**Skills Review 17, p. 23**

1. a. 6,967,284,678     b. 2,517,228,812

2. a. <u>234</u>    6    9    ⑧    3
   b. <u>176</u>    8    4    2    ⑦
   c. <u>330</u>    ④    5    6    3
   d. <u>252</u>    3    ⑧    7    4

3. a. Each one got $30.
   b. $(12 \times \$9 + 17 \times \$6) \div 7$

## Skills Review 17, cont.

**4.**

| a. | | b. | |
|---|---|---|---|
| 146<br>52 ) 7 6 0 3<br>-5 2<br>2 4 0<br>-2 0 8<br>3 2 3<br>- 3 1 2<br>1 1 | 146<br>× 5 2<br>─────<br>2 9 2<br>7 3 0 0<br>─────<br>7 5 9 2<br>+ 1 1<br>─────<br>7 6 0 3 | 114<br>36 ) 4 1 2 7<br>-3 6<br>5 2<br>- 3 6<br>1 6 7<br>- 1 4 4<br>2 3 | 114<br>× 3 6<br>─────<br>6 8 4<br>3 4 2 0<br>─────<br>4 1 0 4<br>+ 2 3<br>─────<br>4 1 2 7 |

## Skills Review 18, p. 24

**1.**

| | |
|---|---|
| a. One way is to round to the nearest hundred. Mary lives in Florida and pays $1,587 ≈ $1,600 a month in rent. Carl lives in Michigan and pays $1,398 ≈ $1,400 a month in rent. Mary pays about __$2,400__ more in rent than Carl in a year. | b. One way is to round to the nearest million, and another is to round to the nearest hundred thousand:<br><br>The states of Wisconsin and Iowa have a population of 5,795,483 ≈ 6,000,000 OR 5,800,000 and 3,145,711 ≈ 3,000,000 OR 3,100,000. The two states have approximately __9,000,000__ OR __8,900,000__ people in total. |

**2.**

Flower 1: center 6; petals 328, 432, 574, 841, 638, 156
Flower 2: center 4; petals 168, 360, 231, 552, 720, 487
Flower 3: center 7; petals 56, 371, 691, 42, 79, 163

**3.** There were 489 × 2 × 2 = 1,956 people who bought tickets to attend.

**4.** Sarah got 27 full rows of photographs. 164 ÷ 6 = 27 R2

**5.** Continue the pattern, counting — until you reach one million.

| |
|---|
| 890,000 |
| 915,000 |
| 940,000 |
| 965,000 |
| 990,000 |
| 1,015,000 |

**6.** In any addition, the numbers that are being added are called __addends__ and the result is called a __sum__ .

In any subtraction, the number you subtract from is called the __minuend__ , the number you subtract is called the __subtrahend__ , and the result is called the __difference__ .

**7.** a. 71    b. 65

1.

| a. My estimation: $50 − 7 × $5 = $15<br>   Exact answer: $17.10 | b. My estimation: $3 × $6 + 2 × $3 = $24<br>   Exact answer: $24.45 |

2. The other side measured 128 ÷ 8 = 16 ft.

3. a. $121 = 11 × 11$     b. $33 = 3 × 11$     c. $84 = 2 × 2 × 3 × 7$

4. a. 5 billion + 3 million + 80 million + 600 billion = <u>605,083,000,000</u>

    b. 20 thousand + 9 million + 4 thousand + 700 million + 12 billion = <u>12,709,024,000</u>

5. $8,000 − 2,320 − 1,890 = x$
  OR  $8,000 − x − 2,320 = 1,890$
  OR  $8,000 − x − 1,890 = 2,320$
  $x + 2,320 + 1,890 = 8,000$
  Solution: $x = 3,790$

# Chapter 3: Problem Solving

## Skills Review 20, p. 26

1. a. 952,798     b. 91,086     c. 676,992

2. a. About 4 hours; mental math     b. About 13 hours; calculator

3.

| | | | | |
|---|---|---|---|---|
| 87 | 95 | 53 | 42 | 123 |
| 153 | 151 | 97 | 167 | 174 |
| 179 | 163 | 109 | 193 | 71 |
| 133 | 127 | 41 | 89 | 195 |
| 106 | 183 | 199 | 161 | 147 |

4.

| | | | | |
|---|---|---|---|---|
| 24,600 | 13,400 | 19,500 | 22,500 | = 80,000 |
| 9,500 | 36,100 | 21,300 | 13,100 | = 80,000 |
| 18,200 | 17,800 | 23,900 | 20,100 | = 80,000 |
| 27,700 | 12,700 | 15,300 | | |
| = 80,000 | = 80,000 | = 80,000 | | |

5.

| | |
|---|---|
| a.  1 million − 300 thousand = 700,000 | b.  1 million − 30 thousand = 970,000 |

## Skills Review 21, p. 27

1.

| a. $23 = 9 + x$ | b. $16 + 38 = 2x$ |
|---|---|
| $x = 14$ | $x = 27$ |

2. a. Yes, because 468 divided by 12 is an even division.
   b. No. Eight doesn't divide evenly into 15, nor into 3,402, so it also doesn't divide evenly into 15 × 3,402.
   c. No. Five doesn't divide evenly into 6, so it also cannot divide evenly into 6 × 6 × 6 × 6 × 6.

3. a. 30,000,000;  8,000,000,000
   b. 60,000;  9,000
   c. 4,700;  23,500,000,000

4. She can print 57 complete copies of the pamphlet.

5. a. 70,072,890    b. 753,859,700

## Skills Review 22, p. 28

1.

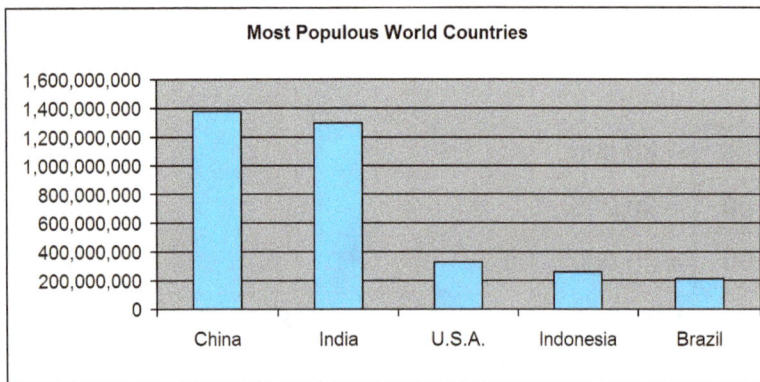

2. a. $238 + 4x = 678$;   $x = 110$
   b. $170 + 2x = 320$;   $x = 75$

3. Each person paid $115.

4. a. 2 × 3 × 3 × 3    b. 2 × 2 × 19    c. 2 × 19

# Chapter 4: Decimals, Part 1

## Skills Review 23, p. 29

1. $420 \div N = 6$;   $N = 70$

2. There were 425 grams of pie left.

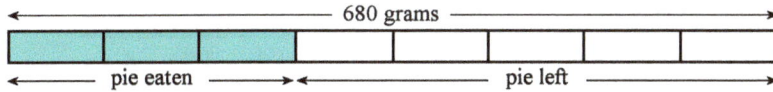

3. 270, 410, 550, 690, 830, 970, 1,110, 1,250, 1,390

4. a. 1,350    b. 1,000    c. 352

5. They raised $2,133.

6. There were 23 shells on each shelf.

7. a. $<$    b. $>$    c. $>$

## Skills Review 24, p. 30

1. a. $6 \times 48 = 6 \times 40 + 6 \times 8 = 240 + 48 = 288$
   b. $9 \times 94 = 9 \times 90 + 9 \times 4 = 810 + 36 = 846$

2. a. 9   b. 51

3.

| Bag 1,328 peaches, 7 peaches in each bag. | | |
|---|---|---|
| **Peaches** | **Bags** | $\begin{array}{r} 189 \\ \hline 7)\,1328 \\ -7 \phantom{000} \\ \hline 62 \phantom{00} \\ -56 \phantom{00} \\ \hline 68 \phantom{0} \\ -63 \phantom{0} \\ \hline 5 \end{array}$ |
| 1328 | | |
| $-700$ | 100 | |
| 628 | | |
| $-560$ | 80 | |
| 68 | | |
| $-63$ | 9 | |
| 5 | 189 | |

4. a. $0.7 + 0.5 = 1.2$
   b. $1.3 - 0.8 = 0.5$

5. The other piece measured 1 ft 10 in.

6. Estimates will vary.

| a. My estimation: $9 + 3 \times 14 = \$51$ | b. My estimation: $\$30 + 2 \times \$20 = \$70$ |
|---|---|
| Exact answer: $\$8.69 + 3 \times \$13.58 = \$49.43$ | Exact answer: $\$28.47 + 2 \times \$19.99 = \$68.45$ |

**Skills Review 25, p. 31**

1. a. 343,215    b. 4,041,038    c. 104,400    d. 6,956,100

2. b.  $3 \times \$15 + \$68$

3. a. 0.083    b. 5.17    c. 8.072    d. 0.3

4.

5.

| Kinds of Flowers | Roses | Daisies | Lilies |
|---|---|---|---|
| Total Number of Flowers | 155 | 147 | 134 |
| Flowers in Each Bouquet | 12 | 15 | 8 |
| Number of Bouquets | 12 | 9 | 16 |
| Flowers Left Over | 11 | 12 | 6 |

**Skills Review 26, p. 32**

1.

a.
```
        2 1 1
  35 ) 7 4 1 0
      7 0
      4 1
    - 3 5
        6 0
      - 3 5
        2 5
```
```
      2 1 1
   ×    3 5
    1 0 5 5
    6 3 3 0
    ───────
    7 3 8 5
  +     2 5
    ───────
    7 4 1 0
```

b.
```
        3 2 3
  17 ) 5 4 9 2
      -5 1
        3 9
      - 3 4
          5 2
        - 5 1
            1
```
```
      3 2 3
   ×    1 7
    2 2 6 1
    3 2 3 0
    ───────
    5 4 9 1
  +       1
    ───────
    5 4 9 2
```

2. a. 4.085,  4.58,  4.8,  4.805,  4.85,  48.5
   b. 9.03,  9.033,  9.3,  9.303,  9.33,  93.03

3.

> $4x + 9 = 57$ | next, remove 9 from both sides
>
> $\quad 4x = 48$
>
> $\quad\; x = 12$

4. a. 1, 2, 3, 6, 19, 38, 57
   b. 1, 2, 41, 82

## Skills Review 27, p. 33

1.

| | |
|---|---|
| a. six squared $= 6^2 = 36$ | d. zero to the seventh power $= 0^7 = 0$ |
| b. $3 \times 3 \times 3 = 3^3 = 27$ | e. $2 \times 2 \times 2 \times 2 = 2^4 = 16$ |
| c. ten to the sixth power $= 10^6 = 1,000,000$ | f. five cubed $= 5^3 = 125$ |

2. They would have 16,068 bones in total.

3. On average, 37 babies are delivered per day.
   On average, 1,133 babies are delivered per month.

4.

| Round to the nearest ... | 12.836 | 0.294 | 5.372 | 29.197 | 8.648 | 36.739 |
|---|---|---|---|---|---|---|
| ... one | 13 | 0 | 5 | 29 | 9 | 37 |
| ... tenth | 12.8 | 0.3 | 5.4 | 29.2 | 8.6 | 36.7 |
| ... hundredth | 12.84 | 0.29 | 5.37 | 29.20 | 8.65 | 36.74 |

5.

| | | | | | | | |
|---|---|---|---|---|---|---|---|
| 24 | 52 | 100 | 403 | 34 | 315 | 855 | 915 |
| 72 | 90 | 66 | 413 | 235 | 613 | 726 | 858 |
| 16 | 102 | 75 | 279 | 498 | 552 | 576 | 743 |
| 63 | 138 | 258 | 144 | 444 | 449 | 568 | 590 |
| 41 | 53 | 294 | 342 | 402 | 486 | 572 | 588 |
| 12 | 27 | 318 | 356 | 309 | 312 | 78 | 544 |
| 22 | 19 | 127 | 98 | 259 | 337 | 348 | 548 |
| 18 | 47 | 66 | 168 | 237 | 394 | 428 | 444 |

6. a. 260    b. 75    c. 162

## Skills Review 28, p. 34

1.

| | fraction/ mixed number | read as ... |
|---|---|---|
| a. 0.796 | 796/1000 | seven hundred ninety six thousandths |
| b. 5.34 | 5 34/100 | five and thirty-four hundredths |

2. $4 \times \$12 + 3x = 87$;  $x = \$13$

3. a. 182    b. - c. Answers will vary. Please check the student's answers.

4. a. 0.94    b. 0.972    c. 0.563

## Skills Review 29, p. 35

1. a. 3.6    b. 0.36    c. 0.036
   d. 2.1    e. 0.21    f. 0.021
   g. 6      h. 6       i. 6

2. The book has about 102,555 words in total.

3. a. 26    b. 183    c. 13,000

4. Answers will vary. Please check the student's answers.

## Skills Review 29, cont.

5.

| a. | | | b. | | |
|---|---|---|---|---|---|
| 116<br>74)8625<br>-74<br>122<br>-74<br>485<br>-444<br>41 | 116<br>× 74<br>464<br>8120<br>8584<br>+ 41<br>8625 | | 393<br>23)9046<br>-69<br>214<br>-207<br>76<br>-69<br>7 | 393<br>× 23<br>1179<br>7860<br>9039<br>+ 7<br>9046 | |

6. a. $x = 2,800,000$   b. $x = 3,500,000$

## Skills Review 30, p. 36

1. 483,786,926

2. a. 62/1000   b. 74 305/1000   c. 8 49/1000   d. 12 7/10

3. Answers will vary, depending on the student's rounding. Please check the student's answers. Below are a few examples:

   a. Rounding to the nearest ten million, the two states had approximately $30,000,000 + 40,000,000 = 70,000,000$ people.

      Rounding to the nearest million, the two states had approximately $28,000,000 + 40,000,000 = 68,000,000$ people.

      Rounding to the nearest hundred thousand, the two states had approximately $28,300,000 + 39,500,000 = 67,800,000$ people.

   b. Rounding the amount of money to the nearest hundred billion and the population to the nearest hundred million, the cost per person is about $900,000,000,000 \div 300,000,000 = \$3,000$.

      Rounding the amount of money to the nearest ten billion and the population to the nearest hundred million, the cost per person is about $870,000,000,000 \div 300,000,000 = \$2,900$.

      Rounding the amount of money to the nearest ten billion and the population to the nearest ten million, the cost per person is about $870,000,000,000 \div 310,000,000 = \$2,806.45$.

4. a. 102.8   b. 24.608   c. $13.72

5. Maya gave 352 haircuts.   $748 - 44 = 704$;   $704 \div 2 = 352$

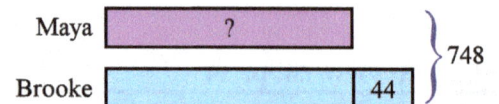

## Skills Review 31, p. 37

1. a.
```
      1
   460.000
   541.320
 +  78.601
 1,079.921
```

   b.
```
   14 16 13
 1 ✗6 ✗10
 ✗5✗.✗0
 - 69.53
  187.87
```

2. Divide $2347 \div 15$ to get 156 R7. Therefore, she can buy 156 ft of fence.

3. There are six different subtractions that are possible to write from the bar model. The student is only required to write one of them. In the addition equation, the addends can be written in any order.

   $91,740 - 18,920 - 30,260 = x$;   $91,740 - 30,260 - 18,920 = x$
   OR $91,740 - x - 18,920 = 30,260$;   OR $91,740 - x - 30,260 = 18,920$
   OR $91,740 - 30,260 - x = 18,920$;   OR $91,740 - 18,920 - x = 30,260$

   $18,920 + x + 30,260 = 91,740$

   Solution: $x = 42,560$

## Skills Review 31, cont.

4. a. Shortcut for repeated addition is multiplication. $80 + 80 + 80 + 80 + 80 + 80$ is $6 \times 80$.
   b. You would need to add 80 __12,500 times__ in order to reach 1 million.
   c. You would need to add 250 __4,000 times__ in order to reach 1 million.
   d. You would need to add 8,000 __125 times__ in order to reach 1 million.

5. a. >   b. =   c. <   d. <

## Skills Review 32, p. 38

1.

Bag 1,572 kiwis, 8 kiwis in each bag.

| Kiwis | Bags |
|-------|------|
| 1572  |      |
| − 800 | 100  |
| 772   |      |
| − 720 | 90   |
| 52    |      |
| − 48  | 6    |
| 4     | 196  |

```
      1 9 6
8) 1 5 7 2
   - 8
     7 7
    - 7 2
       5 2
      - 4 8
         4
```

2.

```
        3 2 4
26 ) 8 4 3 0
    -7 8
      6 3
     - 5 2
       1 1 0
      -1 0 4
           6
```

```
      3 2 4
  ×     2 6
  1 9 4 4
  6 4 8 0
  8 4 2 4
  +     6
  8 4 3 0
```

3. a. 24,000,000   b. 270,000,000   c. 5,600,000,000

4. a. 150   b. 73

5.

| a. 8,49**8**,258,175 | b. **7**3,825,901,637 |
|---|---|
| Place: __one millions place__ | Place: __ten billions place__ |
| Value: __eight million__ | Value: __seventy billion__ |

6. The cheese weighed 6.38 lb in total.

7. a. 0.072   b. 5.6   c. 0.27

## Skills Review 33, p. 39

1. Estimates may vary. Please check the student's answers.
   a. Estimate: $3 + 671 + 5,000 = 5,674$
      Exact: $3.215 + 671.3 + 5,129 = 5,803.515$
   b. Estimate: $81 − 5 = 76$
      Exact: $81.63 − 5.307 = 76.323$
   c. Estimate: $20 \times 7 = 140$
      Exact: $19 \times 7.04 = 133.76$

2.

$$32 = 4x$$
$$8 = x$$
$$x = 8$$

3.

| 10 | 7 | 13 | 5 |
|----|----|----|----|
| 15 | 12 | 28 | 17 |
| 9 | 17 | 4 | 13 |
| 7 | 13 | 42 | 3 |
| 11 | 18 | 15 | 8 |
| 21 | 16 | 2 | 14 |
| 22 | 16 | 6 | 10 |

4.

| Round this to the nearest → | unit (one) | tenth | hundredth |
|---|---|---|---|
| 6.748 | 7 | 6.7 | 6.75 |
| 2.354 | 2 | 2.4 | 2.35 |

| Round this to the nearest → | unit (one) | tenth | hundredth |
|---|---|---|---|
| 14.091 | 14 | 14.1 | 14.09 |
| 11.628 | 12 | 11.6 | 11.63 |

5. a. $7 \times 4 \times T = 140$

b. The expression is $28,000 \div 700$. If we want to find its value (which was not asked but many students may do), we would get 40.

## Skills Review 34, p. 40

1. They would weigh about 146 pounds. $4,657 \div 32 = 145.53125 \approx 146$

2. a. false   b. true   c. false   When changing one number in (a) and (c) answers will vary, as there are several possibilities. For example:

a. $6 + \frac{72}{9} = 14$;  OR   $6 + \frac{63}{9} = 13$;  OR   $5 + \frac{72}{9} = 13$

c. $96 \div 8 = 9 \times 2 - 6$;  OR   $96 \div 8 = 8 \times 2 - 4$

3. a. There were about 2,590 gallons of milk produced per cow in one year. $24,000,000,000 \div 9,267,000 = 2,589.8 \approx 2,590$

b. There were about 7 gallons of milk produced per cow each day. $2,590 \div 365 = 7.095 \approx 7.1 \approx 7$

4. He would need 58 cows. $150,000 \div 2,590 = 57.9 \approx 58$

5. a. $>$   b. $<$

6.

| a. $43 \div 5 = 8$ R3 | b. $77 \div 8 = 9$ R5 |
|---|---|
| $43.0 \div 5 = 8.6$ | $77.000 \div 8 = 9.625$ |
| Check: $5 \times 8.6 = 43.0$ | Check: $8 \times 9.625 = 77.000$ |

# Chapter 5: Statistics and Graphing

## Skills Review 35, p. 41

1. 337,727

2.

| number | 382,526,910 | 7,038,751,328 |
|---|---|---|
| to the nearest 1,000 | 382,527,000 | 7,038,751,000 |
| to the nearest 10,000 | 382,530,000 | 7,038,750,000 |
| to the nearest 100,000 | 382,500,000 | 7,038,800,000 |
| to the nearest million | 383,000,000 | 7,039,000,000 |

3. They had 1,140 points in total. Randy and Lisa had a total of 340 + 340 + 80 = 760 points.
   Alex had half that amount, or 760 ÷ 2 = 380 points. So, in total, they had 760 + 380 = 1,140 points.

4.

1.296   1.316   1.35   1.374   1.408

5.

$11 \times (7 + 9 + 3)$

$= 11 \times 7 + 11 \times 9 + 11 \times 3$

$= 77 + 99 + 33$

$= 209$

6.

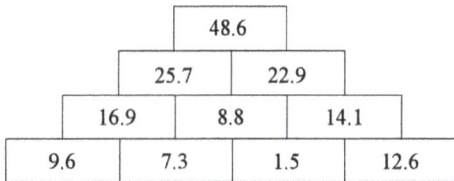

7. $2 \times 2 \times 2 \times 17$

## Skills Review 36, p. 42

1. Price of bicycles

| Bicycles | Price |
|---|---|
| 1 | $178 |
| 2 | $356 |
| 3 | $534 |
| 4 | $712 |
| 5 | $890 |

Capacity of buses of the same size

| Bus | Passengers |
|---|---|
| 1 | 72 |
| 2 | 144 |
| 3 | 216 |
| 4 | 288 |
| 5 | 360 |

2.

743,000,000
823,000,000
903,000,000
983,000,000
1,063,000,000
1,143,000,000
1,223,000,000
1,303,000,000
1,383,000,000
1,463,000,000

Each difference is

80,000,000

## Skills Review 36, cont.

3.

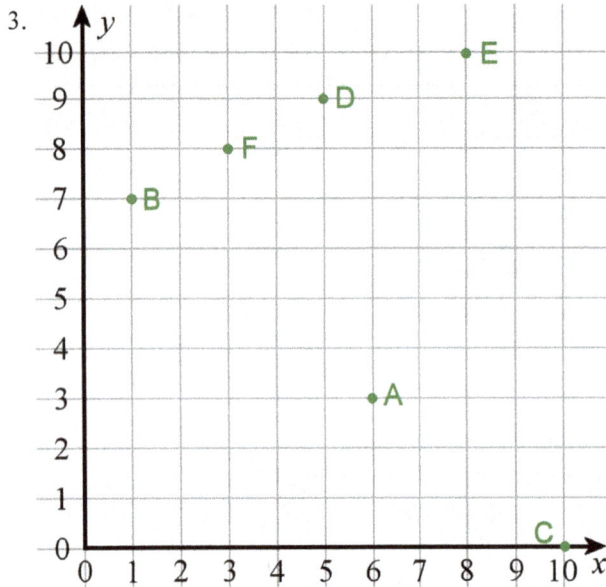

4. a. 4    b. 0.063    c. 0.72

## Skills Review 37, p. 43

1. a. $9/100 + 5/10 + 2/1000 = 0.592$
   b. $13/1000 + 8 + 7/10 = 8.713$

2. Answers will vary. For example:
   a. Approximately how much money have the four friends saved in total?
      Answer: They have saved approximately $2,930 in total.
   b. Approximately how much more money do the four friends have to save
      if they want to share the cost of buying a used car that costs $5,680?
      Answer: They have to save approximately $2,750 more.

3.

$4x + 80 = 136$

| $x$ | $x$ | $x$ | $x$ | 80 |
|-----|-----|-----|-----|-----|

← ——————— 136 ——————— →

$x = 14$
(First subtract 80 from 136. Then divide the result by 4.)

4. a. 42.4    b. 19.25    c. 310.5

## Skills Review 38, p. 44

1.

| $x$ | 10 | 8 | 6 | 4 | 2 | 0 |
|-----|----|----|----|----|----|----|
| $y$ | 5 | 4 | 3 | 2 | 1 | 0 |

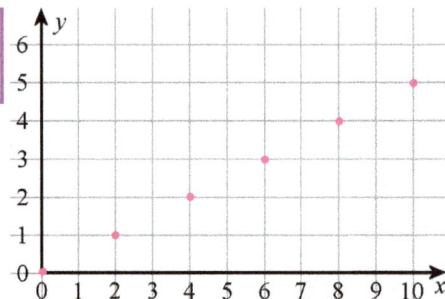

## Skills Review 38, cont.

2.

| a. 7 × (50 + 90) ÷ 4 = 245 | b. 338 = 512 − 6 × (4 + 25) | c. 8 × 5 = 37 + (6 − 5) × 3 |
|---|---|---|

3.

| Divisible by | 2 | 3 | 5 | 6 | 9 |
|---|---|---|---|---|---|
| 756 | x | x | | x | x |
| 960 | x | x | x | x | |

| Divisible by | 2 | 3 | 5 | 6 | 9 |
|---|---|---|---|---|---|
| 5,688 | x | x | | x | x |
| 2,745 | | x | x | | x |

4.

| Amount earned | $70 | $45 | $260 |
|---|---|---|---|
| Amount Carl receives | 42 | 27 | 156 |
| Amount Alec receives | 28 | 18 | 104 |

5. 558,000

6. a. 74/1000     b. 9 623/1000

7. a. 0.08     b. 0.142     c. 7.021     d. 2.63

## Skills Review 39, p. 45

1.

| Month | Total Expenses | Rounded to the Nearest 50 |
|---|---|---|
| Jun | 1,168 | 1,150 |
| Jul | 1,117 | 1,100 |
| Aug | 1,371 | 1,350 |
| Sept | 1,182 | 1,200 |
| Oct | 1,274 | 1,250 |
| Nov | 1,353 | 1,350 |
| Dec | 1,463 | 1,450 |

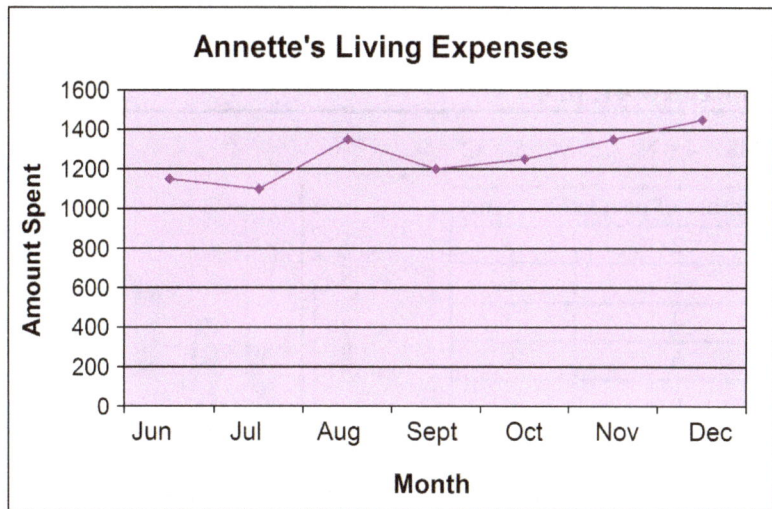

Annette's Living Expenses

2. a. 7     b. 1     c. 7     d. 3

3. Each person blew up 13 balloons, and then some person or people had to blow up 3 more balloons.

4. a. 0.16     b. 0.8     c. 1.5

5. The restaurant can seat 188 people.

6. a. 608     b. 704     c. 509

## Skills Review 40, p. 46

1. a.

| | Liam | Toby |
|---|---|---|
| **Game 1** | 310 | 280 |
| **Game 2** | 260 | 290 |
| **Game 3** | 150 | 220 |
| **Game 4** | 340 | 240 |
| **Game 5** | 190 | 210 |

**Liam and Toby's Game Scores**

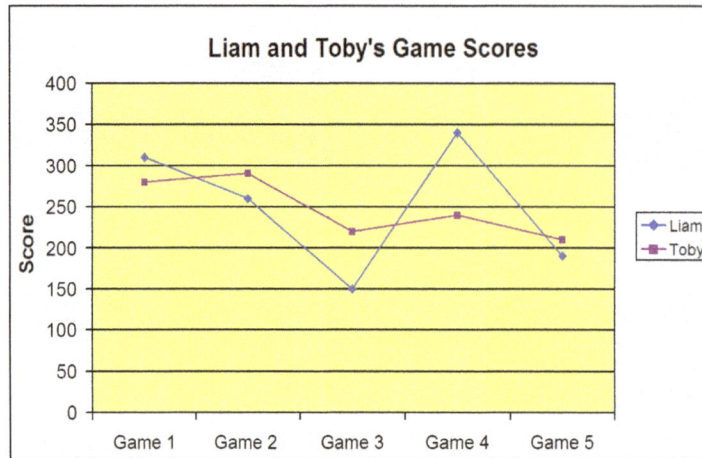

b. Yes, the difference varied a lot. For example, in Game 2, the difference between their scores was 30 points, whereas in Game 4, the difference between their scores was 100 points. Also, sometimes Liam won, and sometimes Toby won.

2. About 16 hours.

3. The approximate amount spent on medicine per person was $1,170.

4. a. 70,000    b. 500,000,000    c. 240,000,000

Mystery Number:  280;  369

## Skills Review 41, p. 47

1. a. $148 \div 4 = N$    b. $M \div 8 = 12$

2.

| Number of people | Frequency |
|---|---|
| 2 | 3 |
| 3 | 4 |
| 4 | 5 |
| 5 | 5 |
| 6 | 2 |
| 7 | 2 |
| 8 | 1 |

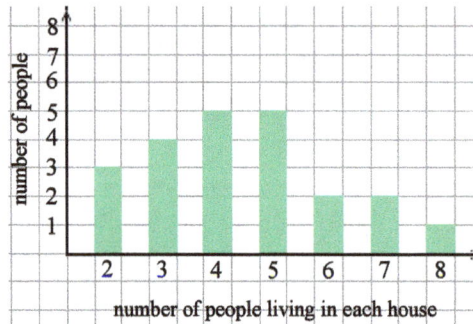

number of people living in each house

3. a. $37 + 37 + 2x = 190$; $x = 58$ cm.
   b. $\$15 + 6x = \$63$;  $x = \$8$

4.

| a. 146 × 0.4 | b. 9 × 6.072 | c. 15 × 8.91 |
|---|---|---|

a. 146 × 0.4

```
      1 2
    1 4 6
  ×   0.4
    5 8.4
```

b. 9 × 6.072

```
        6 1
    6.0 7 2
  ×       9
  5 4.6 4 8
```

c. 15 × 8.91

```
        4
      8.9 1
  ×     1 5
    4 4 5 5
    8 9 1 0
  1 3 3.6 5
```

## Skills Review 42, p. 48

1. a. 0.006    b. 0.06    c. 0.9

2. Answers will vary, because when drawing a histogram, the bin width can vary slightly,
   and so can the starting point of the first bin. The bin width is $(72 - 5) \div 5 = 13.4 \approx 14$ units.
   The bins can start at 3, 4, or 5 years. In the graph below, the first bin starts at 5.

| age | frequency |
|-----|-----------|
| 5-18 | 6 |
| 19-32 | 3 |
| 33-46 | 2 |
| 47-60 | 3 |
| 61-74 | 2 |

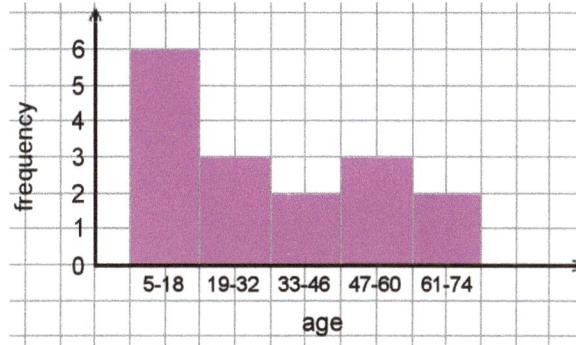

3. a. $2 \times 2 \times 2 \times 2 \times 3 \times 3$
   b. $2 \times 2 \times 2 \times 3 \times 5$
   c. $2 \times 2 \times 2 \times 11$

4. Each one paid $3,566.50.

5. Answers may vary, depending on the student's rounding. For example:

> The cities of Las Vegas and Reno have $632,912 \approx 630,000$ and $245,255 \approx 250,000$ population.
> The two cities have approximately  880,000  people in all.
> There are about  380,000  more people in Las Vegas than in Reno.

## Skills Review 43, p. 49

1. a.

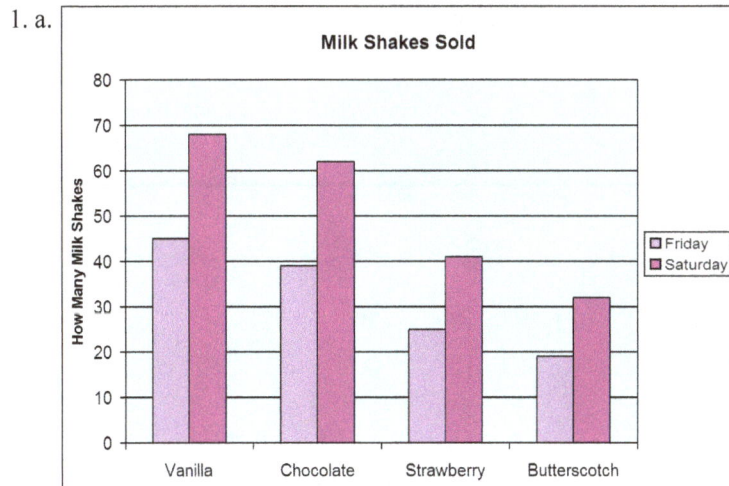

   b. Chocolate had the most difference in sales, with 23 more sold on
      Saturday than on Friday.

2. a. 49,530,004    b. 386,863,241

3. a. 339,700    b. 468,733

4.

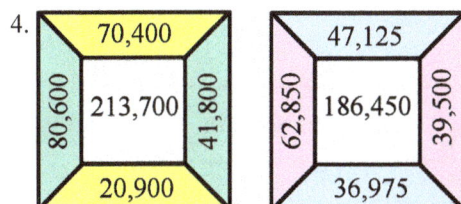

# Chapter 6: Decimals, Part 2

## Skills Review 44, p. 50

1. a. 6.59  Check: $14 \times 6.59 = 92.26$
   b. 2.419  Check $37 \times 2.419 = 89.503$

2. a. $240 \div (4 \times 15) - 4 = 0$
   b. $76 + 9 \times 8 \div 12 = 82$

3. a. 664.2   b. 18.09   c. 119.28

4. a. >   b. >   c. <

5. a. 0.6   b. 0.3   c. 0.8   d. 0.4

6.

| 820 | 1,640 | 3,280 | 6,560 | 13,120 | 26,240 | 52,480 |
|-----|-------|-------|-------|--------|--------|--------|

## Skills Review 45, p. 51

1.

> My estimation: $70 - 9 \times \$7 = \$7$
> Exact answer: $5.20

2. $5{,}728 \times 60 = 343{,}680$

3. One block is $348 \div 6 = \$58$.
   Brett earned $5 \times \$58 = \$290$.

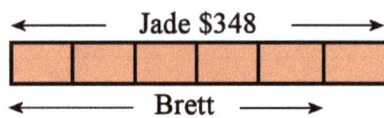

4. a. >   b. <   c. >   d. >

5. It will take him 36 days. $(256 + 378) \div 18 = 35 \text{ R}4$

6. a. 0.58   b. 0.407   c. 0.099   d. 0.016

## Skills Review 46, p. 52

1. a. 0.56   b. 1.5   c. 0.48   d. 1   e. 0.105   f. 4.2

2. a.

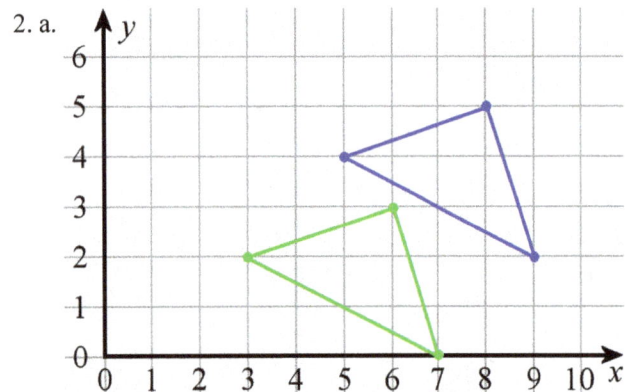

   b. $(3, 2)$, $(7, 0)$, $(6, 3)$

3.

| Round this to the nearest → | unit (one) | tenth | hundredth | Round this to the nearest → | unit (one) | tenth | hundredth |
|-----|-----|-----|-----|-----|-----|-----|-----|
| 3 . 8 6 2 | 4 | 3.9 | 3.86 | 9 . 5 3 7 | 10 | 9.5 | 9.54 |

## Skills Review 46, cont.

4. a. 5.5    b. 517.5

5.

| a. | Check: | b. | Check: |
|---|---|---|---|

a.
```
          1 3 5
    6 9 ) 9 3 1 8
        6 9
        2 4 1
        2 0 7
            3 4 8
            3 4 5
                3
```
Check:
```
      1 3 5
    ×   6 9
    ─────────
    1 2 1 5
    8 1 0 0
    ─────────
    9 3 1 5
    +     3
    ─────────
    9 3 1 8
```

b.
```
            1 3 3
    3 5 ) 4 6 7 3
        3 5
        1 1 7
        1 0 5
            1 2 3
            1 0 5
                1 8
```
Check:
```
      1 3 3
    ×   3 5
    ─────────
      6 6 5
    3 9 9 0
    ─────────
    4 6 5 5
    +   1 8
    ─────────
    4 6 7 3
```

Puzzle Corner:  a. 127, 131, 137, 139    b. 96;  The 12 factors are: 1, 2, 3, 4, 6, 8, 12, 16, 24, 32, 48, 96.

## Skills Review 47, p. 54

1. a. "Blue."

b.

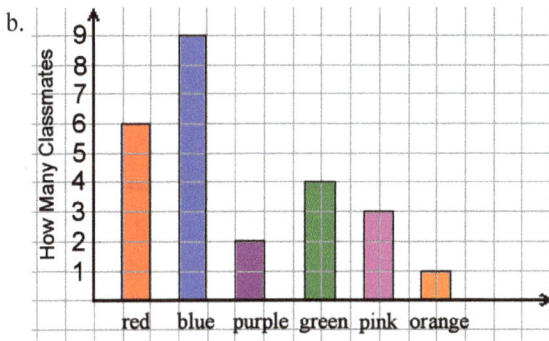

c. It isn't possible, because the data is not numerical.

2. a. 20 thousand + 30 million + 500 million + 90 billion = 90,530,020,000
   b. 70 million + 5 billion + 2 thousand + 6 million + 400 million = 5,476,002,000

3.

| a. $21 + 47 = 4x$ | b. $4x = 6 + 9 + x$ |
|---|---|
| $4x = 68$ | $4x = 15 + x$ |
| $x = 17$ | $x = 5$ |

4. a. $0.824 = 0.8 + 0.02 + 0.004$    b. $0.092 = 0.09 + 0.002$

5. $\underline{8} + 9 = 17$       difference

  $\underline{21} - 6 = 15$       subtrahend

  $37 + 5 = \underline{42}$       sum

  $15 - \underline{6} = 9$       minuend

  $43 - 7 = \underline{36}$       addend

6.

| a. $2.5 × 60$ px $= 150$ px | b. $0.8 × 60$ px $= 48$ px |
|---|---|

## Skills Review 48, p. 55

1. a. You can get $3.2 \div 0.2 = \underline{16}$ pieces.
   b. You can get $5.6 \div 0.8 = \underline{7}$ pieces.

2.

| x | 5 | 6 | 7 | 8 | 9 | 10 |
|---|---|---|---|---|---|----|
| y | 1 | 2 | 3 | 4 | 5 | 6  |

The rule is: $y = x - 4$

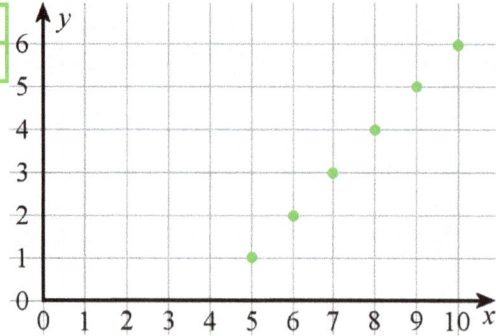

3. a. 22.63    b. 80.7

4. a. 8,560,374,239    b. 797,406,942

5.
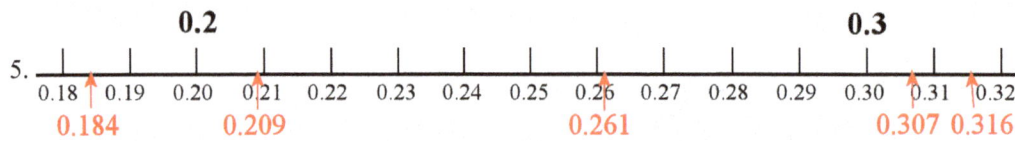

6. a. $9 \times 10^5 = 900,000$   |   b. $5 \times 10^7 = 50,000,000$   |   c. $73 \times 10^9 = 73,000,000,000$

## Skills Review 49, p. 56

1. a.

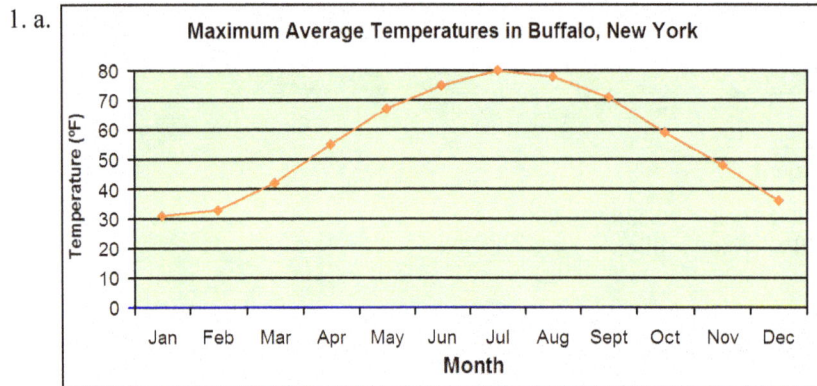

Maximum Average Temperatures in Buffalo, New York

   b. The average of the average temperatures is 56.3 °F.

2. a. 400;  0.009    b. 0.06;  37    c. 0.59;  0.423

3. There are 198.1 grams in seven ounces.

4. When you divide one tenth into ten equal parts, you get <u>hundredths</u>.
   When you divide one hundredth into ten equal parts, you get <u>thousandths</u>.

5.

   a. The simplest way is to round the amount of money spent to the nearest thousand, and round the number of weeks in a year to the nearest ten. The Carson family spent $14,765 ≈ $15,000 last year on groceries. They spent about $15,000 ÷ 50 = $300 weekly.

   b. The best way is to round the number of miles to the nearest hundred. Al drives 1,080 ≈ 1,100 miles round-trip to work monthly. He drives about $1,100 \times 12 = 13,200$ miles to work yearly.

6. $154 = 2 \times 7 \times 11$

## Skills Review 50, p. 57

1. a. Approximately 1,470,000 new cars and trucks were sold per month.
   b. Some students may decide to take the answer from a. and divide it by 30, which will produce an incorrect answer; since not all months have 30 days. The correct calculation would be 17,600,000 ÷ 365 = 48,219.17808219. Then, round that answer to the nearest hundred: approximately 48,200 new cars and trucks were sold per day.

2. a.

| Day | Number of people | Rounded to the nearest 1000 |
|-----|------------------|------------------------------|
| Wed | 8,922 | 9,000 |
| Thu | 12,275 | 12,000 |
| Fri | 17,351 | 17,000 |
| Sat | 28,936 | 29,000 |
| Sun | 36,550 | 37,000 |

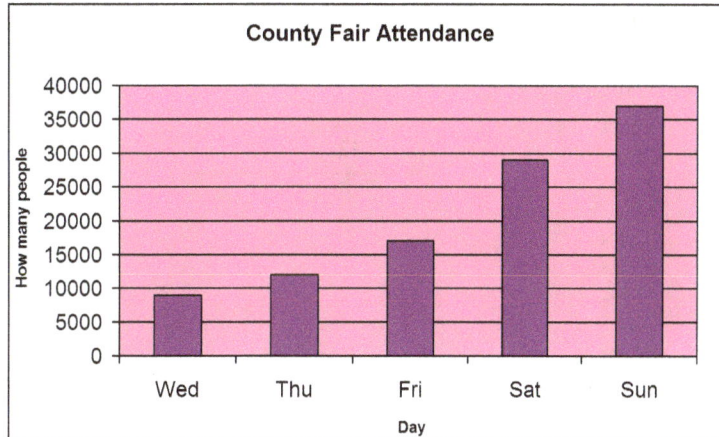

**County Fair Attendance**

b. On average, about 20,800 people attended each day.

3.

```
        1 6 9
   27 ) 4 5 8 3
      - 2 7
        1 8 8
      - 1 6 2
          2 6 3
        - 2 4 3
            2 0
```

```
        1 6 9
      ×   2 7
      _____
      1 1 8 3
      3 3 8 0
      _____
      4 5 6 3
    +     2 0
      _____
      4 5 8 3
```

4. Estimate: $0.8 \times 3 = 2.4$     Exact: $0.8 \times 2.64 = 2.112$

## Skills Review 51, p. 58

1.

$\frac{4}{7} - 4 \div 7 = 0.571$

```
        0 . 5 7 1 4
   7 ) 4 . 0 0 0 0
       3 5
         5 0
       - 4 9
           1 0
         -  7
             3 0
           - 2 8
               2
```

2. a. 322     b. 292

3. a. 900 ml; 5,400 ml     b. 7.43 L; 3.015 L     c. 72 ml; 2,650 ml

4. a. 3     b. 8

31

5. Answers will vary, because when drawing a histogram, the bin width can vary slightly, and so can the starting point of the first bin. The bin width can be $(48 - 4) \div 5 = 8.8 \approx 9$ units. In this case, the bins will start at 4, 13, 22, 31, and 40. A bin with of 10 will work also. The important part is that the smallest data item should fit in the first bin, and the largest in the last bin. The graph shows the histogram with a bin width of 9.

| number of miles | frequency |
|---|---|
| 4-12 | 11 |
| 13-21 | 7 |
| 22-30 | 2 |
| 31-39 | 3 |
| 40-48 | 2 |

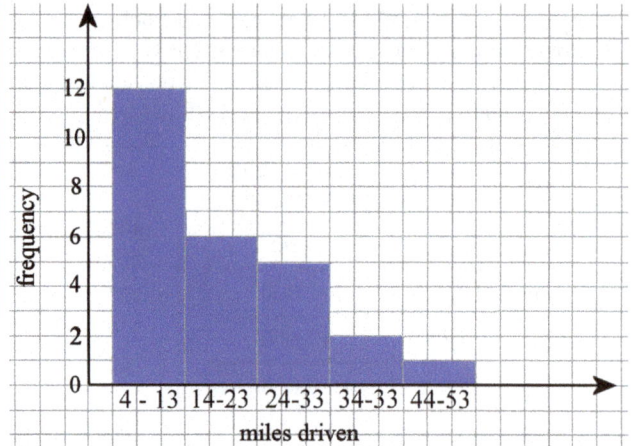

**Skills Review 52, p. 59**

1. a. 60 oz;   3 lb 15 oz
   b. 144 oz;   2 lb 6 oz
   c. 4 lb 8 oz;   3 lb 2 oz

2. a. A number is divisible by  10  if it ends in 0.
   b. A number is divisible by  6  if it is divisible by both 2 and 3.
   c. A number is divisible by  8  if half of it is divisible by 4.
   d. A number is divisible by  5  if it ends in 0 or 5.

3.

$4x + 8 = 56$ | next, remove 8 from both sides

$\quad 4x = 48$

$\quad\quad x = 12$

4.

Bag 939 lemons, 8 lemons in each bag.

| Lemons | Bags |
|---|---|
| 9 3 9 | |
| − 8 0 0 | 1 0 0 |
| 1 3 9 | |
| −   8 0 | 1 0 |
| 5 9 | |
| −  5 6 | 7 |
| 3 | 1 1 7 |

```
      1 1 7
8 ) 9 3 9
    -8
    1 3
    - 8
      5 9
    - 5 6
        3
```

5.

$4.768 \div 0.08 = \underline{59.6}$

$47.68 \div 0.8$

$476.8 \div 8$

```
        5 9 . 6
8 ) 4 7 6 . 8
  - 4 0
      7 6
    - 7 2
        4 8
      - 4 8
          0
```

Check:

```
    5 9 . 6
  ×       8
  4 7 6 . 8
```

# Chapter 7: Fractions: Add and Subtract

## Skills Review 53, p. 60

1.

| a. $8^7 > 1,000,000$ | b. $0^{\square} > 1,000,000$    impossible | c. $23^5 > 1,000,000$ |
|---|---|---|

2. There was 0.39 lb of ham left. First, find one tenth of the package of ham: $1.3 \text{ lb} \div 10 = 0.13 \text{ lb}$.
   Now, multiply that by 7 to find how much ham was used: $0.13 \text{ lb} \times 7 = 0.91 \text{ lb}$.
   Lastly, subtract the amount used from the original amount: $1.3 \text{ lb} - 0.91 \text{ lb} = 0.39 \text{ lb}$.

3. a. The average weight is 3.3 lb.
   b. Now the average weight is 3.1. The average decreased by $3.3 - 3.1 = 0.2$ lb.

4. a. 728 m ≈ 1 km; rounding error = 0.272 km
   b. 14.516 km ≈ 15 km; rounding error = 0.484 km
   c. 8.399 km ≈ 8 km; rounding error = 0.399 km

5. a. $3.14 \div 0.009 = 348.9$. Transform $3.14 \div 0.009$ into $31.4 \div 0.09$, then into $314 \div 0.9$, and lastly into $3140 \div 9$. Remember to use 3140.00 in the long division, so you can get the answer to 2 decimal digits, and then round it to 1 decimal digit.
   b. $57 \div 0.36 = 158.33$. Transform $57 \div 0.36$ into $570 \div 3.6$ and then into $5700 \div 36$. Use 5700.000 in the long division, so you can get the answer to 3 decimal digits, and then round it to 2 decimal digits.

6.

| a.6. | 0 | 5 | 3 |  | b.5. |
|---|---|---|---|---|---|
| 0 |  |  |  |  | 0 |
| 4 |  |  | c.9. | 0 | 4 |
|  | d.0. | 0 | 2 |  | 8 |
|  | 0 |  |  |  |  |
|  | 0 |  | e.0. | 9 | 8 |
| f.0. | 5 | 3 | 7 |  |  |

## Skills Review 54, p. 61

1. a. The mode is Spanish.

   b. **Languages Studied**

| Language | Frequency |
|---|---|
| Chinese | 3 |
| Italian | 4 |
| French | 12 |
| German | 7 |
| Spanish | 18 |

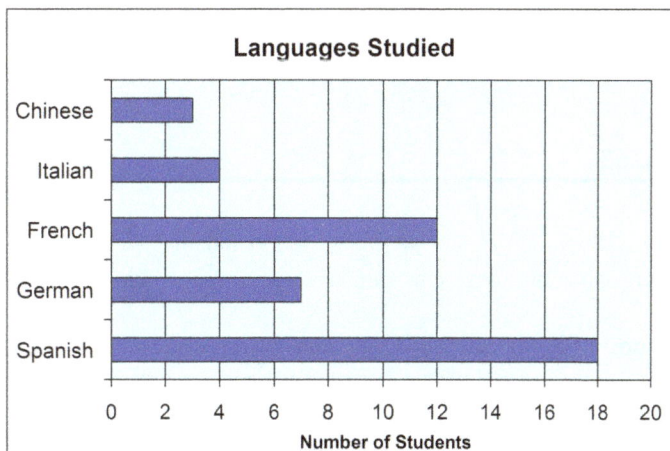

Languages Studied

   c. It isn't possible.    d. There were 44 students in all, and $12/44 = 3/11$ of them studied French.

2. a. 0.752, 2.75, 5.72, 7.52, 57.2, 75.2     b. 3.678, 3.867, 38.617, 38.671, 38.761, 386.71

3. a. <u>Equivalent</u> fractions are equal in value.     b. <u>Unlike</u> fractions have a different denominator.
   c. A <u>proper</u> fraction is a fraction that is less than one.     d. <u>Like</u> fractions have the same denominator.

4. They paid $27,636 in total.

5. $60 ÷ 8 = $7.50; $7.50 × 7 = $52.50  OR $60 − $7.50 = $52.50.
The new price is $52.50.

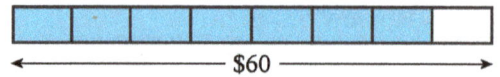

## Skills Review 55, p. 62

1. Estimates will vary. Estimate: 4,000 × 70 = 280,000;  Exact: 274,032

2. a. 8   b. 9 2/5   c. 9 4/7

3. a. 12   b. 0.09   c. 11   d. 70   e. 0.8   f. 0.11

4.

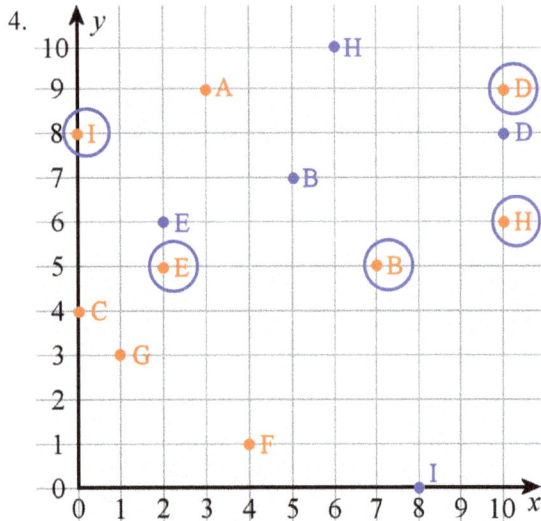

5. In any multiplication, the numbers that are multiplied are called _factors_ and the
result is called a _product_ . Sixty-three is divisible by both 7 and 9. In other
words, 7 and 9 are _divisors_ of 63.

6.

| Round this to the nearest → | unit (one) | tenth | hundredth |
|---|---|---|---|
| 1 2 . 5 9 6 | 13 | 12.6 | 12.60 |

| Round this to the nearest → | unit (one) | tenth | hundredth |
|---|---|---|---|
| 8 . 0 8 2 | 8 | 8.1 | 8.08 |

## Skills Review 56, p. 63

1. a. 643.9 m

| 6 | 4 | 3. | 9 | | | |
|---|---|---|---|---|---|---|
| km | hm | dam | m | dm | cm | mm |

b. 391.8 dam

| 3 | 9 | 1. | 8 | | | |
|---|---|---|---|---|---|---|
| km | hm | dam | m | dm | cm | mm |

c. 643.9 m = _6,439_ dm = _64,390_ cm = _643,900_ mm

391.8 dam = _3,918_ m = _39,180_ dm = _391,800_ cm

2. a. 2 2/9   b. 6/7   c. 1

3. $236,762 ÷ $46 = 5,147 tickets sold.

4. a. 0.572   b. 7.06   c. 3.7

## Skills Review 56, cont.

5. a.

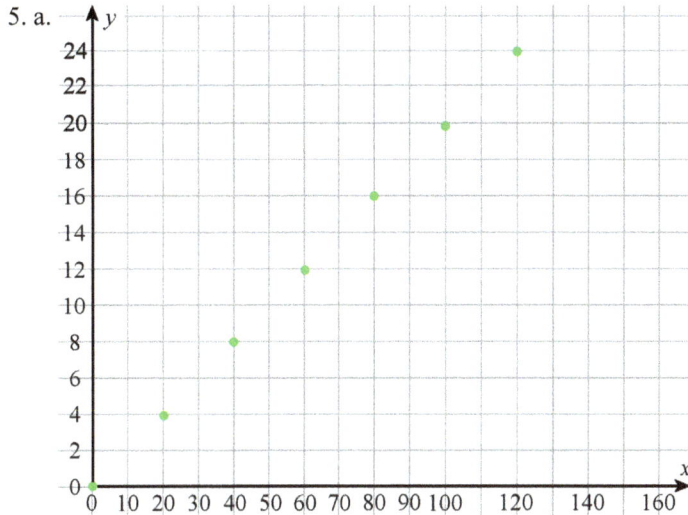

b. $y = \dfrac{x}{5}$

| x | 0 | 20 | 40 | 60 | 80 | 100 | 120 |
|---|---|----|----|----|----|-----|-----|
| y | 0 | 4  | 8  | 12 | 16 | 20  | 24  |

## Skills Review 57, p. 64

1. There are six different subtractions that are possible to write from the bar model. The student is only required to write one of them. In the addition equation, the addends can be written in any order.

   $300 - 47 - 110 = x$
   OR $300 - 110 - 47 = x$
   OR $300 - x - 110 = 47$
   OR $300 - x - 47 = 110$
   OR $300 - 110 - x = 47$
   OR $300 - 47 - x = 110$
   $47 + 110 + x = 300$
   Solution: $x = 143$

2. a. 600   b. 4,000,000   c. 2,800,000

3. Sarah ate $0.045 \div 3 \times 2 = 0.03$ kg of chocolate.
   Don ate $0.2 \div 4 = 0.05$ kg of chocolate.
   Don ate 0.02 kg more chocolate than Sarah.

4. a. 0.063   b. 0.66   c. 0.88

5. a. Answers may vary, depending on the student's estimates. Rounding to the nearest whole ten, they answered about 220 e-mails.
   b. Paula answered the most e-mails during the five days.

6. a. 12   b. 9   c. 4

35

## Skills Review 58, p. 65

1.
```
      1 1
   5 2 6 . 3 0 0
     1 8 . 4 0 0
 +      3 . 9 0 7
 ─────────────────
   5 4 8 . 6 0 7
```

2. 1, 2, 4, 7, 8, 14, 16, 28, 56, 112

3. a. 2 6/8    b. 8 8/12

4. a.

| State | Population | Rounded to the nearest 100,000 |
|-------|-----------|-------------------------------|
| Ohio | 11,658,609 | 11,700,000 |
| Idaho | 1,716,943 | 1,700,000 |
| Georgia | 10,429,379 | 10,400,000 |
| Indiana | 6,666,818 | 6,700,000 |
| Virginia | 8,470,020 | 8,500,000 |

b.
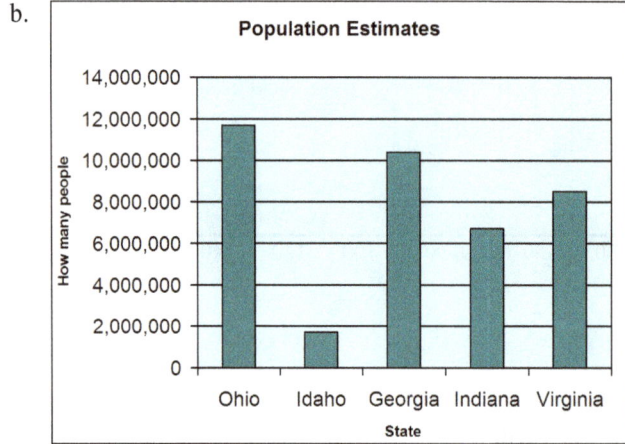

5. a. 9 C 1 oz    b. 10 C    c. 9 gal 3 qt

## Skills Review 59, p. 66

1.
$$\frac{7}{12} = 7 \div 12 = 0.5833 \approx 0.583$$

```
          0 . 5 8 3 3
    12 ) 7 . 0 0 0 0
        -6 0
        ─────
         1 0 0
        -  9 6
        ──────
             4 0
           - 3 6
           ──────
               4 0
             - 3 6
             ──────
                 4
```

2. a. 1 2/3    b. 5 4/6    c. 8/2    d. 3 5/15

3. 262,907,980

4.

| a. 2.4 kg − 530 g | b. 1.7 L − 362 ml |
|---|---|
| = 1.87 kg ≈ <u>1.9 kg</u> | = 1.338 L ≈ <u>1.3 L</u> |

5. $9 \times (9 + 12)$

$= 9 \times 9 + 9 \times 12$

$= 81 + 108 = 189$

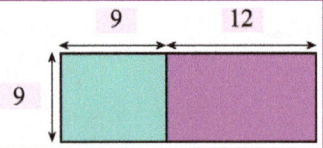

36

## Skills Review 60, p. 67

1.

| a.  176 <br> 41 ⟌ 7 2 3 9 <br> -4 1 <br> 3 1 3 <br> -2 8 7 <br> 2 6 9 <br> -2 4 6 <br> 2 3 | 1 7 6 <br> × 4 1 <br> 1 7 6 <br> +7 0 4 0 <br> 7 2 1 6 <br> + 2 3 <br> 7 2 3 9 | b.  2 3.7 <br> 26 ⟌ 6 1 8.5 <br> -5 2 <br> 9 8 <br> - 7 8 <br> 2 0 5 <br> -1 8 2 <br> 2 3 | 2 3.7 <br> × 2 6 <br> 1 4 2 2 <br> +4 7 4 0 <br> 6 1 6 2 <br> + 2 3 <br> 6 1 8.5 |

2. a. 1.08    b. 0.04    c. 0.021

3. a. The coloring of the numbers 4, 6, 12, and 14 corresponds with the coloring of the prime numbers which are their prime factors. For example, the number 4 has two blue-colored sections, and since the number 2 is colored blue, this illustrates that the number 4's prime factors are 2 × 2.

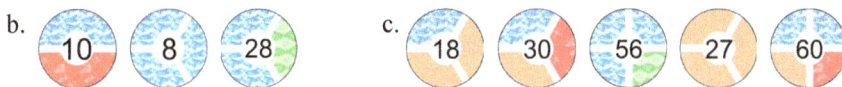

b.    10    8    28        c.    18    30    56    27    60

4.

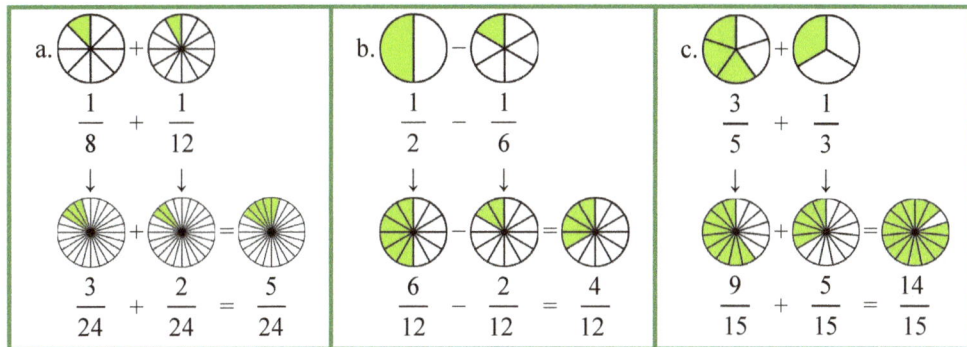

| a.  $\frac{1}{8} + \frac{1}{12}$ <br> ↓    ↓ <br> $\frac{3}{24} + \frac{2}{24} = \frac{5}{24}$ | b.  $\frac{1}{2} - \frac{1}{6}$ <br> ↓    ↓ <br> $\frac{6}{12} - \frac{2}{12} = \frac{4}{12}$ | c.  $\frac{3}{5} + \frac{1}{3}$ <br> ↓    ↓ <br> $\frac{9}{15} + \frac{5}{15} = \frac{14}{15}$ |

## Skills Review 61, p. 68

1. a.

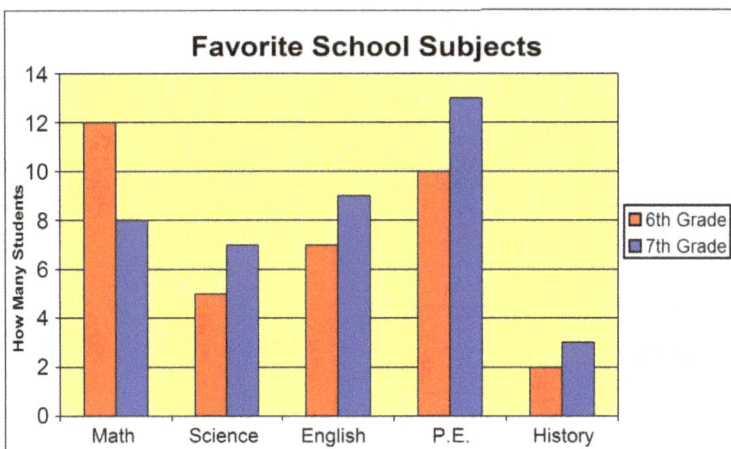

**Favorite School Subjects**

b. The original data consists of *words*, such as "P.E., math, math, English, history, English...".
   Therefore, it is not possible to calculate an average for the data.

2. a. About 331,541 live births per month.        b. About 76,510 live births per week.

3. Estimate: 1 × 4.28 = 4.28  OR 0.8 × 4 = 3.2  OR  0.9 × 4 = 3.6
   Exact:    0.8 × 4.28 = 3.424

4. a. 0.004    b. 0.062

37

## Skills Review 61, cont.

**5.**

| fractions to add/subtract | c. d. |
|---|---|
| a.  5th parts  and  9th parts | 45 |
| b.  4th parts  and  3rd parts | 12 |

**6.**

a. $\dfrac{2}{5} + \dfrac{1}{9}$    $\downarrow$    $\downarrow$    $\dfrac{18}{45} + \dfrac{5}{45} = \dfrac{23}{45}$

b. $\dfrac{3}{4} - \dfrac{2}{3}$    $\downarrow$    $\downarrow$    $\dfrac{9}{12} - \dfrac{8}{12} = \dfrac{1}{12}$

## Skills Review 62, p. 69

1. a. 6.8    b. 736.5

2. a. $6\frac{3}{8} \Rightarrow 6\frac{15}{40}$
$\quad -\ 4\frac{1}{10} \quad -\ 4\frac{4}{40}$
$\quad\quad\quad\quad\quad\quad 2\frac{11}{40}$

b. $7\frac{3}{4} \Rightarrow 7\frac{9}{12}$
$\quad +\ 4\frac{5}{6} \quad +\ 4\frac{10}{12}$
$\quad\quad\quad\quad 11\frac{19}{12} \to 12\frac{7}{12}$

3. a. 299.2 oz    b. 12,860 lb    c. 5.75 lb

4. a. 95.2    b. 92.222    c. 68.57    d. 8.2

5. a. 17.5 cm = _1.75_ dm = _0.175_ m = _0.0175_ dam

| | | | | 1 | 7. | 5 |
|---|---|---|---|---|---|---|
| km | hm | dam | m | dm | cm | mm |

b. 9.6 km = _96_ hm = _960_ dam = _9,600_ m

| 9. | 6 | | | | | |
|---|---|---|---|---|---|---|
| km | hm | dam | m | dm | cm | mm |

c. 24.19 hm = _241.9_ dam = _2,419_ m = _24,190_ dm

| 2 | 4. | 1 | 9 | | | |
|---|---|---|---|---|---|---|
| km | hm | dam | m | dm | cm | mm |

6.

| number | 96,319,540 | 512,851,274 |
|---|---|---|
| to the nearest 10,000 | 96,320,000 | 512,850,000 |
| to the nearest 100,000 | 96,300,000 | 512,900,000 |
| to the nearest million | 96,000,000 | 513,000,000 |

7. a. 0.4    b. 0.4    c. 0.09

# Chapter 8: Fractions: Multiply and Divide

## Skills Review 63, p. 70

1. a. 36/63 > 35/63    b. 16/24 < 18/24

2.

> My estimation: 40 × $4 + 3 × $30 = $250
> Exact answer: 42 × $3.84 + 3 × $31.95 = $257.13

3. a.  $0.6 \times \underline{\hspace{1cm}} = \underline{\hspace{1cm}}$

      $0.6 \times 40$ px = $\underline{24}$ px

  b.  $2.3 \times \underline{\hspace{1cm}} = \underline{\hspace{1cm}}$

      $2.3 \times 40$ px = $\underline{92}$ px

4. Aaron and Emily ate 9/24 + 4/24 = 13/24 of the pie, so 11/24 of the pie was left.

5. a.

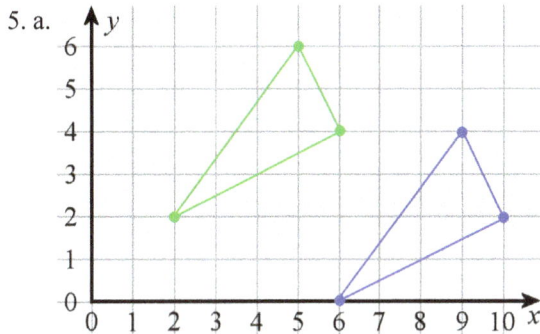

  b.  (5, 6),  (2, 2),  (6, 4)

6.

| a. 2.3 m − 74 cm = 1.56 m ≈ 1.6 m | b. 5.25 m + 34 cm + 3.9 m = 9.49 m ≈ 9.5 m | c. 6 m 9 cm + 9.42 m = 15.51 m ≈ 15.5 m |
|---|---|---|

## Skills Review 64, p. 71

1. a. 93    b. 12

2. a. 4 2/3    b. 1/8    c. 3/8    d. 6 3/4

3.

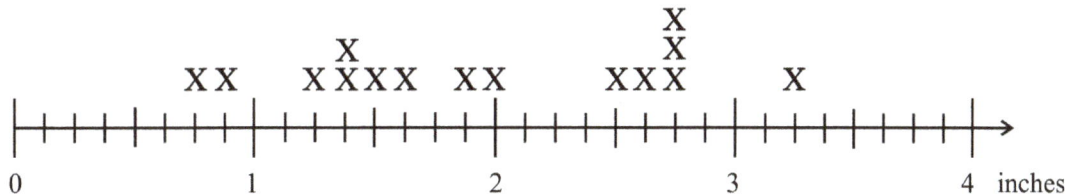

4. a. >    b. <    c. =    d. >

5. 7 kg ÷ 0.45 kg = 15.6 lb;  She ended up with 15 full bags of flour.

6.

| a. $8 \times 10^{8} = 800{,}000{,}000$ | b. $5 \times 10^{6} = 5{,}000{,}000$ | c. $72 \times 10^{4} = 720{,}000$ |
|---|---|---|

Puzzle Corner:  a. 1,260 or 1,620 or 2,160 or 6,120    b. 6,201 or 2,601 or 6,021 or 2,061

## Skills Review 65, p. 71

1. a.  $3 \times 18/20 = 3/1 \times 9/10 = 27/10$ or $2\ 7/10$     b. $13/7 \times 5 = 13/7 \times 5/1 = 65/7$ or $9\ 2/7$

2. a. 50,309,020,000    b. 603,500,064,000

3.

| | | a. 3 | | d. 0. | 5 |
|---|---|---|---|---|---|
| | a. 2 | 0 | | 0 | |
| | | | c. 1 | 2 | |
| b. 0. | 0 | 0 | 4 | | e. 0. |
| 0 | | | | e. 1 | 0 |
| 2 | | | | | 3 |

4. a. 48/5   b. 103/9   c. 79/12   d. 19/4

5. a. Kate spent the most time talking on the phone.     b. June.

## Skills Review 66, p. 73

1.

| Round this to the nearest → | unit (one) | tenth | hundredth |
|---|---|---|---|
| 6.247 | 6 | 6.2 | 6.25 |
| 8.561 | 9 | 8.6 | 8.56 |
| 2.309 | 2 | 2.3 | 2.31 |
| 4.825 | 5 | 4.8 | 4.83 |

2.

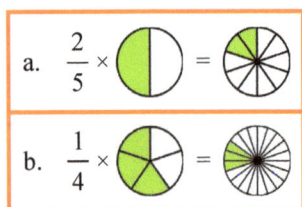

a. $\dfrac{2}{5} \times$   = 

b. $\dfrac{1}{4} \times$   = 

3.

$x + 5 + 24 = 3x + 13$ | next, remove $x$ from both sides

$5 + 24 = 2x + 13$

$29 = 2x + 13$

$x = 8$

4. a. 3.21 Check: $27 \times 3.21 = 86.67$
   b. 4.534 Check: $14 \times 4.534 = 63.476$

5. The two parts in the model that are the same size are $4,320 − $480 = $3,840 in total. One of those parts is therefore $3,840 ÷ 2 = $1,920. Marla earns $1,920, and Nikki earns $1,920 + $480 = $2,400.

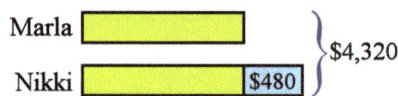

## Skills Review 67, p. 74

1.

| a. $9 \frac{6}{7}$ | b. $10 \frac{11}{12}$ |
|---|---|
| $5 \frac{2}{7}$ | $8 \frac{4}{12}$ |
| $+\ 7 \frac{5}{7}$ | $+\ 6 \frac{7}{12}$ |
| $21 \frac{13}{7} \rightarrow 22 \frac{6}{7}$ | $24 \frac{22}{12} \rightarrow 25 \frac{10}{12}$ |

2. In each problem, the factors may also be written in the other order.

a. $\dfrac{4}{5} \times \dfrac{3}{4} = \dfrac{12}{20}$ 　　　 b. $\dfrac{2}{5} \times \dfrac{3}{3} = \dfrac{6}{15}$

3. a. - c. Answers will vary. Please check the student's answers.

4.

|  | a. 725 cm | b. 4,230 mm | c. 8.3 dm |
|---|---|---|---|
| meters | 7.25 | 4.23 | 0.83 |
| decimeters | 72.5 | 42.3 | 8.3 |
| centimeters | 725 | 423 | 83 |
| millimeters | 7,250 | 4,230 | 830 |

5. He had 1.02 miles left to run. The distance he had already run was $7/10 \times 3.4$ miles $= 0.7 \times 3.4$ miles $= 2.38$ miles. The distance he had left was $3.4$ miles $- 2.38$ miles $= 1.02$ miles.

6. a. 48　 b. 0.7　 c. 0.2

## Skills Review 68, p. 75

1.

a.

| O |  | t | h | th |
|---|---|---|---|---|
| 0 | . | 8 | 1 | 4 |

$= \dfrac{814}{1000}$

$= 8 \times \dfrac{1}{10} + 1 \times \dfrac{1}{100} + 4 \times \dfrac{1}{1000}$

b.

| O |  | t | h | th |
|---|---|---|---|---|
| 0 | . | 7 | 2 | 5 |

$= \dfrac{725}{1000}$

$= 7 \times \dfrac{1}{10} + 2 \times \dfrac{1}{100} + 5 \times \dfrac{1}{1000}$

2. a. 2 5/11　 b. 15 5/8

3. Students don't need to write the intermediate multiplication step. For clarity, it is shown here in its entirety.

a. $\dfrac{\overset{1}{\cancel{3}}}{\underset{4}{\cancel{16}}} \times \dfrac{\overset{5}{\cancel{20}}}{\underset{3}{\cancel{9}}} = \dfrac{1 \times 5}{4 \times 3} = \dfrac{5}{12}$ 　　 b. $\dfrac{\overset{1}{\cancel{3}}}{\underset{1}{4}} \times \dfrac{\overset{4}{\cancel{16}}}{\underset{15}{\cancel{45}}} = \dfrac{1 \times 4}{1 \times 15} = \dfrac{4}{15}$

**Skills Review 68, cont.**

4.

| | | |
|---|---|---|
| a. $2.760 \div 0.8 = \underline{3.45}$ <br><br> $27.60 \div 8$ | $\begin{array}{r} 3.4\,5 \\ 8\,)\overline{2\,7.6\,0} \\ \underline{-2\,4} \\ 3\;6 \\ \underline{-\;3\,2} \\ 4\;0 \\ \underline{-\;4\,0} \\ 0 \end{array}$ | Check: <br><br> $\begin{array}{r} 3.4\,5 \\ \times\quad 8 \\ \hline 2\,7.6\,0 \end{array}$ |
| b. $0.852 \div 0.03 = \underline{28.4}$ <br><br> $8.52 \div 0.3$ <br><br> $85.2 \div 3$ | $\begin{array}{r} 2\;8.4 \\ 3\,)\overline{8\;5.2} \\ \underline{-6} \\ 2\;5 \\ \underline{-2\;4} \\ 1\;2 \\ \underline{-1\;2} \\ 0 \end{array}$ | Check: <br><br> $\begin{array}{r} 2\,8.4 \\ \times\quad 3 \\ \hline 8\,5.2 \end{array}$ |

5. a. 11,950.4 yd   b. 0.15 mi   c. 2,366.67 yd

## Skills Review 69, p. 76

1. a. The average page count was 158 pages.
   b. $241 - 158 = 83$ more pages.

2.

| a. Estimate: $9 \times 4 = 36$ <br> Exact: $9.3 \times 4.12 = 38.316$ | b. Estimate: $8 \times 6 = 48$ <br> Exact: $7.57 \times 6.1 = 46.177$ |
|---|---|

3.

| 63 | 81 | 59 | 45 |
|---|---|---|---|
| 92 | 15 | 43 | 111 |
| 13 | 137 | 101 | 159 |
| 199 | 169 | 87 | 123 |
| 11 | 163 | 51 | 177 |
| 119 | 149 | 67 | 89 |
| 153 | 117 | 141 | 193 |

4. a. 0.005   b. 0.009   c. 0.013   d. 0.017   e. 0.024

5. a. 26 3/7   b. 46 7/12

## Skills Review 70, p. 77

1. a. =   b. >   c. <

2. Each person paid ($80.65 + $89.99) ÷ 4 = $42.66.

3. The original piece weighed 2.75 oz. First, find how much one-fifth of the piece of cheesecake weighed by dividing the weight of the remaining 2/5 of the piece of cheesecake by 2: 1.1 oz ÷ 2 = 0.55 oz. Now, multiply by 5 to get the weight of the original piece of cheesecake: 0.55 oz × 5 = 2.75 oz.

## Skills Review 70, cont.

4.

<table>
<tr>
<td>a. $\dfrac{7}{9} - \dfrac{1}{5}$<br>↓    ↓<br>$\dfrac{35}{45} - \dfrac{9}{45} = \dfrac{26}{45}$</td>
<td>b. $\dfrac{3}{8} + \dfrac{9}{10}$<br>↓    ↓<br>$\dfrac{15}{40} + \dfrac{36}{40} = \dfrac{51}{40} = 1\dfrac{11}{40}$</td>
<td>c. $\dfrac{4}{7} - \dfrac{2}{9}$<br>↓    ↓<br>$\dfrac{36}{63} - \dfrac{14}{63} = \dfrac{22}{63}$</td>
</tr>
</table>

5.

a. 912 mm

| km | hm | dam | m | dm | cm | mm |
|----|----|-----|---|----|----|----|
|    |    |     | 9 | 1  | 2  |    |

b. 8.74 km

| km | hm | dam | m | dm | cm | mm |
|----|----|-----|---|----|----|----|
| 8. | 7  | 4   |   |    |    |    |

c. 912 mm = <u>91.2</u> cm = <u>9.12</u> dm = <u>0.912</u> m

8.74 km = <u>87.4</u> hm = <u>874</u> dam = <u>8,740</u> m

6. a. 0.54   b. 0.72   c. 0.014   d. 0.04   e. 1.28   f. 0.84

## Skills Review 71, p. 78

| 1. a. 2.4 m − 0.78 m = 1.62 m<br>≈ 1.6 m | b. 7.34 m + 0.59 m + 4.8 m = 12.73 m<br>≈ 12.7 m | c. 3.16 m + 7.63 m = 10.79 m<br>≈ 10.8 m |
|---|---|---|

2. a. 9/24, 4/6, 8/27, and 13/38 are incorrect. Answers will vary. Possible correct answers are:
2/6, 5/15, 8/24, 9/27, 10/30, 11/33, 12/36, 13/39, 14/42, 15/45, 16/48, 18/54, 19/57, and 20/60.

b. 12/15, 18/24, 24/30, 2/6, and 12/21 are incorrect. Answers will vary. Possible correct answers are:
2/3, 6/9, 10/15, 12/18, 14/21, 16/24, 18/27, 20/30, 24/36, 28/42, 30/45, 32/48, 34/51, 36/54, 38/57, and 40/60.

3. a. About $940 per month.      b. About $220 per week.

4.

| a. Divide these 2 pies equally between five people. Each will get $\dfrac{2}{5}$ of a pie. | b. Divide these 4 pies equally between six people. Each will get $\dfrac{4}{6}$ or $\dfrac{2}{3}$ of a pie. |
|---|---|

5. a. 120 pt;  34 C      b. 13 gal 1 qt;  19 pt      c. 9 C 4 oz;  104 qt

6. Jay will need to work 25 full hours in order to have enough money to buy the canoe. ($369.99 − $74.35) ÷ 12 = $24.64 ≈ $25

**Skills Review 72, p. 79**

1. a. 7 28/63 − 3 45/63 = 3 46/63      b. 4 10/15 + 8 9/15 = 12 19/15 = 13 4/15

2. a. The mode is 13 and 15.

b.

| age | Frequency |
|-----|-----------|
| 8..9 | 3 |
| 10..11 | 5 |
| 12..13 | 7 |
| 14..15 | 6 |
| 16..17 | 4 |

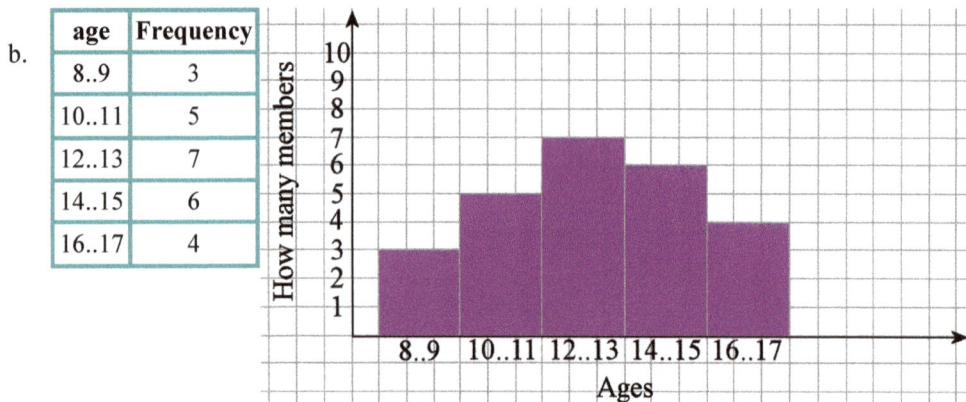

c. The mean is 12.8 ≈ 13.

3. a. 0.105    b. 0.036

4. a. 115 oz    b. 3 lb 13 oz

5.

| a. $\dfrac{1}{4} \div 6$ | b. $9 \div 3$ | c. $\dfrac{1}{9} \div 12$ |
|---|---|---|
| ↓ ↓ ↓ | ↓ ↓ ↓ | ↓ ↓ ↓ |
| $\dfrac{1}{4} \times \dfrac{1}{6} = \dfrac{1}{42}$ | $9 \times \dfrac{1}{3} = 3$ | $\dfrac{1}{9} \times \dfrac{1}{12} - \dfrac{1}{108}$ |

**Skills Review 73, p. 80**

1.

| a. Rounding the amount of money to $1500 and rounding the number of weeks in a year to 50 makes the problem simple.  Seth spent $1,593 ≈ $1,500 on gasoline last year. He spent about $1,500 / 50 = $30 weekly. | b. Answers may vary. Check the student's answer. For example:  Kelly pays about $126.50 ≈ $130 a month for bus fare. She pays about 12 × $130 = $1,560 for bus fare in a year. |
|---|---|

2. a. LCD is 21. 18/21 + 7/21 = 25/21 = 1 4/21
   b. LCD is 36. 28/36 − 15/36 = 13/36
   c. LCD is 88. 33/88 + 16/88 = 49/88

3. a. dekagrams    b. centigrams    c. grams

4. a.
```
    7.2 0 0
    4.3 1 6
+   9.5 9 0
─────────────
  2 1.1 0 6
```
b.
```
  2 5.3 0 0
−    8.5 2 4
─────────────
  3 3.8 2 4
```
c.
```
  1 3.2 0
     6.4 8
+    2.0 7
─────────────
  2 1.7 5
```

| 5. a. $\dfrac{2}{5} < \dfrac{5}{8} < \dfrac{7}{10}$ | b. $\dfrac{5}{12} < \dfrac{3}{4} < \dfrac{7}{9}$ | c. $\dfrac{1}{7} < \dfrac{1}{5} < \dfrac{4}{11}$ |
|---|---|---|

6.

$$\frac{1}{5} \div 3 = \frac{1}{15}$$

Check: $\frac{1}{15} \times 3 = \frac{3}{15} = \frac{1}{5}$

7. a. 1/24   b. 3/11   c. 1/40

# Chapter 9: Geometry

## Skills Review 74, p. 82

1. Please check the student's drawings. The images below are example answers:

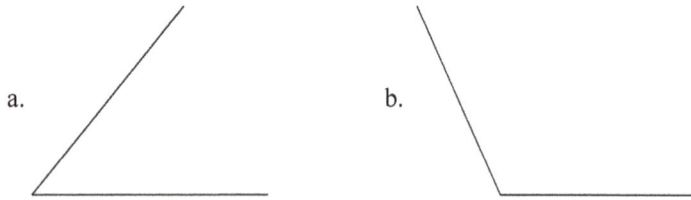

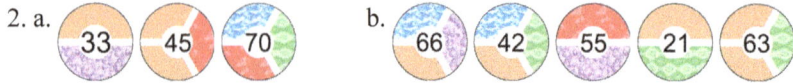

a.

b.

2. a. 33 45 70    b. 66 42 55 21 63

3. a.
$$
\begin{array}{r}
1\ 1\ 1\quad1\\
1\ 8\ 2\ .\ 0\ 0\ 0\\
4\ 7\ 5\ .\ 3\ 8\ 0\\
+\quad2\ 9\ .\ 7\ 3\ 9\\
\hline
6\ 8\ 7\ .\ 1\ 1\ 9
\end{array}
$$

b.
$$
\begin{array}{r}
9\\
1\,10\,13\quad5\,10\\
2\ 0\ 3\ .\ 6\ 0\\
-\quad6\ 7\ .\ 1\ 9\\
\hline
1\ 3\ 6\ .\ 4\ 1
\end{array}
$$

4. 17 3/4 inches

5. a. 56   b. 60

6. a. 0.497   b. 3.805

## Skills Review 75, p. 83

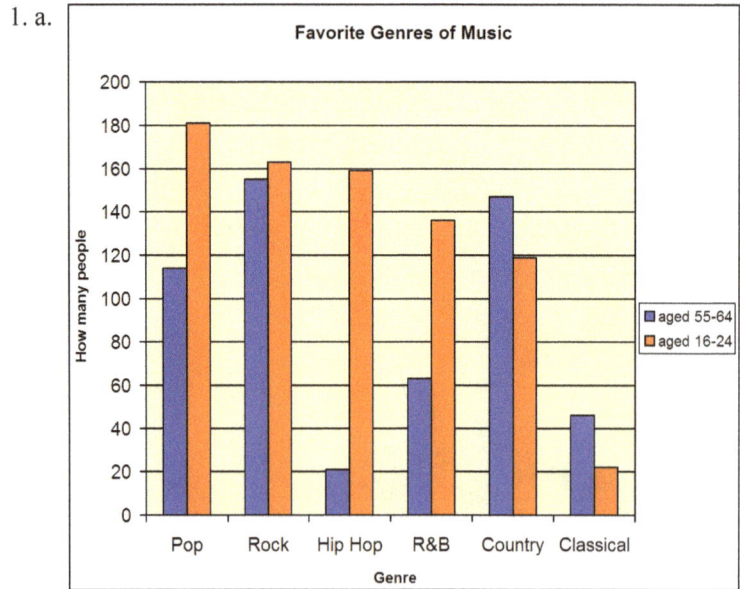

1. a.

**Favorite Genres of Music**

(Bar chart showing "How many people" on the y-axis from 0 to 200 and "Genre" on the x-axis with categories: Pop, Rock, Hip Hop, R&B, Country, Classical. Two series: aged 55-64, aged 16-24.)

b. Rock

2. a. 200   b. 12   c. 60

## Skills Review 75, cont.

3.

| a. $\dfrac{8}{11} - \dfrac{3}{7}$ | b. $\dfrac{1}{4} + \dfrac{2}{9}$ |
|---|---|
| $\downarrow \qquad \downarrow$ | $\downarrow \qquad \downarrow$ |
| $\dfrac{56}{77} - \dfrac{33}{77} = \dfrac{23}{77}$ | $\dfrac{9}{36} + \dfrac{8}{36} = \dfrac{17}{36}$ |

4. a. Check that the student copied the figure accurately.
   b. Triangle A is right. Triangle B is obtuse. Triangle C is right.

## Skills Review 76, p. 84

|  | **18.506** | **53.172** |
|---|---|---|
| one | 19 | 53 |
| tenth | 18.5 | 53.2 |
| hundredth | 18.51 | 53.17 |

2. a. 54/9 = 6
   b. 60/100 = 30/50 = 15/25 = 3/5

3. a. - b. Please check the student's drawings.

4. a. 124.6    b. 54.315    c. 156.06

5. The estimates will vary. The exact measures are:
   a. 29°    b. 82°    c. 120°

6. a. 2 2/13    b. cannot simplify    c. 3 3/5

## Skills Review 77, p. 85

1. a. - b. Answers will vary. Please check the student's drawings.
         An example drawing is on the right.
      c. Three diagonals; four triangles.
      d. Answers will vary. Please check the student's answers.

2. a. 40/63    b. 9/44

3. a. 158.07    b. 123.3    c. 127.143    d. 1.42

4. a. A  trapezoid  has one pair of parallel sides.
   b. A  parallelogram  has two pairs of parallel sides.
   c. A  square  is a rectangle with four congruent sides.
   d. A  kite  has two pairs of congruent sides.

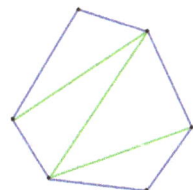

5. a. "apple" and "banana"

b.

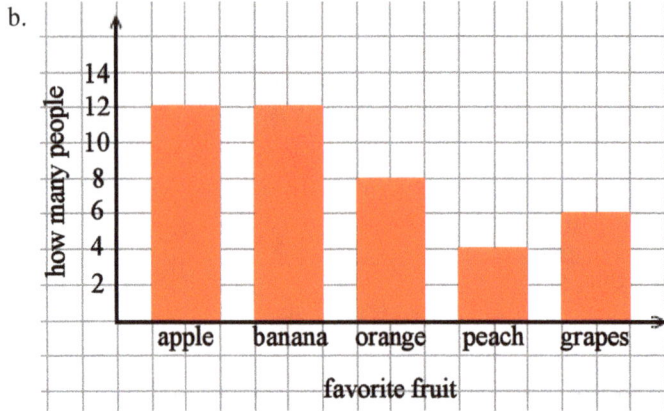

c. It isn't possible.

d. 8/42 = 4/21 of the people liked oranges best.

## Skills Review 78, p. 86

1. a. 380    b. 0.004

2. a. - b.

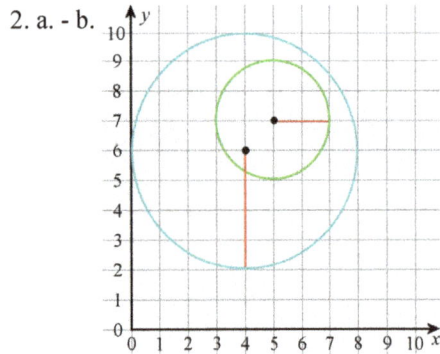

3. a. 1,366.67 yd    b. 4.09 mi    c. 50,688 ft

4. a. isosceles    b. obtuse

5. a. 3 1/21    b. 15 5/24

6.

| | | | | | | | |
|---|---|---|---|---|---|---|---|
| $\frac{4}{10}$ | $\frac{6}{15}$ | $\frac{1}{5}$ | $\frac{3}{6}$ | $\frac{6}{9}$ | $\frac{6}{18}$ | $\frac{5}{15}$ | $\frac{4}{9}$ |
| $\frac{7}{12}$ | $\frac{12}{20}$ | $\frac{10}{25}$ | $\frac{4}{11}$ | $\frac{27}{39}$ | $\frac{10}{15}$ | $\frac{8}{24}$ | $\frac{13}{37}$ |
| $\frac{8}{11}$ | $\frac{17}{24}$ | $\frac{14}{35}$ | $\frac{22}{30}$ | $\frac{8}{12}$ | $\frac{15}{24}$ | $\frac{4}{10}$ | $\frac{11}{33}$ |
| $\frac{20}{55}$ | $\frac{8}{20}$ | $\frac{18}{40}$ | $\frac{10}{14}$ | $\frac{14}{21}$ | $\frac{18}{29}$ | $\frac{7}{21}$ | $\frac{15}{40}$ |
| $\frac{14}{40}$ | $\frac{12}{30}$ | $\frac{20}{50}$ | $\frac{12}{16}$ | $\frac{24}{36}$ | $\frac{14}{42}$ | $\frac{12}{32}$ | $\frac{9}{28}$ |
| $\frac{15}{35}$ | $\frac{30}{70}$ | $\frac{16}{40}$ | $\frac{30}{45}$ | $\frac{34}{46}$ | $\frac{17}{51}$ | $\frac{22}{63}$ | $\frac{16}{54}$ |

## Skills Review 79, p. 88

1. Answers will vary. The images on the right are just examples. Please check the student's drawings.

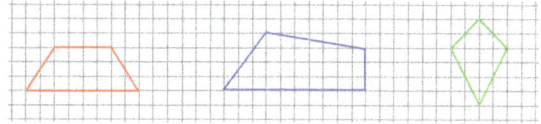

2. The perimeter is 90 ft.  476 ft ÷ 17 ft = 28 ft;
   2 × 17 ft + 2 × 28 ft = 90 ft.

3. a. The grapes cost $3 1/2 × 4 1/2 = ($7/2) × (9/2) = $63/4 = $15 3/4 = $15.75.
      You can also use decimals: $3.50 × 4.5 = $15.75

   b. 4.75 × $3.60 = $17.10

4. 9,772 ÷ 39 = 250 R22

5. a. 0.21    b. 0.024    c. 0.36

6.

| | |
|---|---|
| a. $\dfrac{\overset{1}{\cancel{34}}}{\underset{1}{\cancel{12}}} \times \overset{1}{\cancel{12}} = 34$ | |
| b. $\overset{7}{\cancel{56}} \times \dfrac{9}{\underset{1}{\cancel{8}}} = 63$ | |

## Skills Review 80, p. 89

1. a.

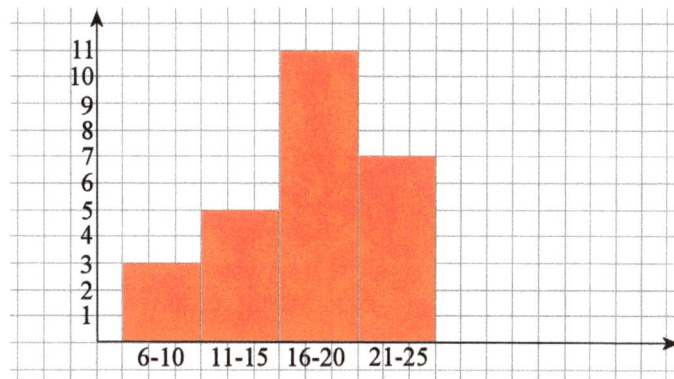

   b. acute equilateral

2. Answers will vary, because the bin width can vary slightly, and so can the starting point of the first bin. The bin width is (25 − 7) ÷ 4 = 4.5 ≈ 5 units. In the graph below, the first bin starts at 6, but it could also start at 5 or 7 and still be correct.

| correct answers | frequency |
|---|---|
| 6 - 10 | 3 |
| 11 - 15 | 5 |
| 16 - 20 | 11 |
| 21 - 25 | 7 |

3. Check the student's answer. The orientation of the prism may vary. The image on the right is one possible answer.

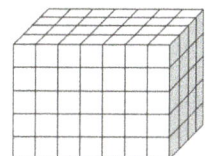

4. The range is: numbers from  275,000  to  284,999 .

5. a. 1 7/16 in   b. 5/8 in

49

## Skills Review 81, p. 90

1. The answers may vary slightly, depending on the student's estimates. Mike drove approximately 178.3 miles per day, and Roger drove approximately 171.7 miles per day.

2.

| | | |
|---|---|---|
| a. | | $\frac{4}{7}$ ft $\times$ $\frac{3}{4}$ ft $=$ $\frac{12}{28}$ ft$^2$ $=$ $\frac{3}{7}$ ft$^2$ |
| b. | | $\frac{2}{3}$ km $\times$ $\frac{8}{9}$ km $=$ $\frac{16}{27}$ km$^2$ |

3. a. 7.002, 7.02, 7.12, 7.2, 70.2, 72.02
   b. 3.416, 3.461, 30.461, 34.061, 34.61, 346.1

4. a. 5/9   b. 3   c. 1/6

5. a. A pair of binoculars costs $48.60. Divide the price of the puzzle by 3 to get 1/5 of the price of the binoculars: $29.16 ÷ 3 = $9.72. Then, multiply by 5: $9.72 × 5 = $48.60.

   b. $262.44;  3 × $48.60 + 4 × $29.16 = $262.44

6. V = 70 ft × 25 ft × 6 ft = 70 ft × 150 ft$^2$ = 10,500 ft$^3$

## Skills Review 82, p. 91

1. 630,712,833

2.

$7^5 =$ _16,807_
$7^6 =$ _117,649_
$7^7 =$ _823,543_
$7^8 =$ _5,764,801_

3. a. $12 3/5
   b. $12.60

4. a. 632 mm

| km | hm | dam | m | dm | cm | mm |
|---|---|---|---|---|---|---|
| | | | | 6 | 3 | 2 |

   b. 5.81 km

| km | hm | dam | m | dm | cm | mm |
|---|---|---|---|---|---|---|
| 5. | 8 | 1 | | | | |

   c. 632 mm = _63.2_ cm = _6.32_ dm = _0.0632_ dam

   5.81 km = _58.1_ hm = _581_ dam = _581,000_ cm

5. 5 cm

Mystery Number: 86;  128

1. Answers will vary. Check the student's answer. For example:

   Tamara cut six tomatoes into fourths. How many pieces did she end up with?
   $6 \div (1/4) = 24$. She ended up with 24 pieces of tomato.

2. The area of the other room is 550 ft². The area of the entire building is 22 ft × 40 ft = 880 ft². Subtract the area of the one room from the total area to get the area of the other room: 880 ft² − 330 ft² = 550 ft².

3. The estimate will vary. The exact measure is 163°.

4. The depth is 4 in.

5. a. - b. Please check the student's drawings.

6.

| x | 1 ½ | 3 | 4 ½ | 6 | 7 ½ | 9 |
|---|-----|---|-----|---|-----|---|
| y | 6 | 5 ½ | 5 | 4 ½ | 4 | 3 ½ |

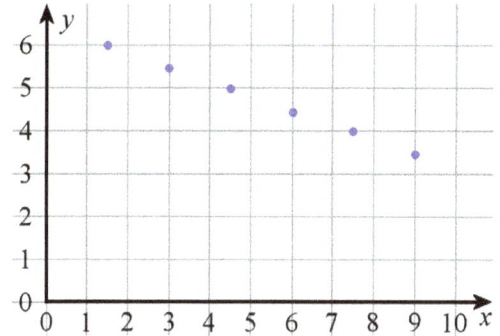

www.ingramcontent.com/pod-product-compliance
Lightning Source LLC
Chambersburg PA
CBHW080722220326
41520CB00056B/7377